ENVIRONMENTAL REPORTING, RECORDKEEPING, AND INSPECTIONS

T0318934

JOHN WILEY & SONS, INC.

New York • Chichester • Weinheim • Brisbane • Singapore • Toronto

This book is printed on acid-free paper. ⊖

This publication is designed to provide accurate and authoritative information in regard to the subject matter covered. It is sold with the understanding that the publisher is not engaged in rendering professional services. If professional advice or other expert assistance is required, the services of a competent professional person should be sought.

Library of Congress Cataloging-in-Publication Data:

Dennison, Mark S.
 Environmental reporting, recordkeeping, and inspections : a compliance guide for business and industry/Mark S. Dennison.
   p. cm.
 Includes bibliographical references and index.
  ISBN 0-471-29074-2
   1. Environmental law—United States. 2. Hazardous substances—Health aspects—Reporting—United States. I. Title.
KF3775.D43 1995
344.73'046—dc20   95-22390
[347.30446]

10 9 8 7 6 5 4 3

For Tracey, whose love has been my guiding light

Contents

Chapter 2
State Environmental Law - Overview

PART II - ENVIRONMENTAL REPORTING AND RECORDKEEPING

Chapter 3
Environmental Reporting and Recordkeeping Requirements

Chapter 4
Environmental Reporting and Recordkeeping Procedures

Chapter 5
Emergency Response / Waste Minimization / Compliance Audits

PART III - ENVIRONMENTAL INSPECTIONS

Chapter 6
Handling Environmental Inspections

Chapter 7
Environmental Inspection Procedures

Preface

Environmental Reporting, Recordkeeping, and Inspections: A Compliance Guide for Business and Industry has been written as an all-purpose guide for complying with the reporting and recordkeeping requirements of the federal and state environmental laws and regulations. The book has been divided into three parts, plus appendices.

Part I provides a helpful introduction to the most important federal and state environmental laws affecting business and industry. Part II examines the specific reporting and recordkeeping requirements and procedures for each of these federal and state environmental laws. This part also offers helpful emergency response planning and waste minimization measures and discusses the importance of performing environmental compliance audits. Finally, Part III shows how to handle environmental inspections and explains the inspection procedures used by government regulators. Helpful inspection checklists round out this part of the book. Practical tables and checklists are used throughout to summarize key points and provide quick reference guides to regulatory requirements. The appendices contain helpful listings of federal and state environmental regulatory agencies and offices.

Considerable effort has been made to write the book in easy-to-understand language, explaining the environmental regulatory requirements in practical terms that a company's environmental manager, attorney, or consultant can apply directly to a business or to a facility's day-to-day operations. It is hoped that this book will function for many years to come as a regulatory compliance guide and practical reference to handling environmental reporting, recordkeeping, and inspections.

Mark S. Dennison
Ridgewood, New Jersey

Acknowledgements

I wish to express my appreciation to everyone who played a role in publication of this book. I thank the editorial and production staff at Van Nostrand Reinhold, including Bob Esposito, Jane Kinney, and Barbara Mathieu, for all their assistance. I especially wish to thank my former editor, Alex Padro, for developing the original idea for the book, and for providing his editorial leadership throughout the publishing process.

I also wish to thank the many helpful individuals working at different federal and state regulatory offices who answered my questions and provided copies of various forms and government documents.

Additional thanks go to family and friends for providing moral support and inspiration, especially Erin Carrather, Jonathan Huff, Jessie Ann Huff, Bobbie Waits, Harry Huff, Jr., Mrs. J. Howard Gould, and Keith Dennison.

But my greatest thanks must go to my dear Tracey, whose constant love and support inspired me to complete this book.

M.S.D.

Acronyms

ARAR	Applicable or Relevant and Appropriate Requirements
BACT	Best Available Control Technology
CAA	Clean Air Act
CERCLA	Comprehensive Environmental Response, Compensation and Liability Act
CFR	Code of Federal Regulations
CWA	Clean Water Act
DOT	Department of Transportation (U.S.)
EA	Environmental Assessment
EIS	Environmental Impact Statement
EHS	Extremely Hazardous Substance
EPA	Environmental Protection Agency
EPCRA	Emergency Planning and Community Right-to-Know Act
FIFRA	Federal Insecticide, Fungicide, and Rodenticide Act
FONSI	Finding of No Significant Impact
GACT	Generally Available Control Technology
HSWA	Hazardous and Solid Waste Amendments of 1984
LDRs	Land Disposal Restrictions
LEPC	Local Emergency Planning Committee
MACT	Maximum Achievable Control Technology
MSDS	Material Safety Data Sheet
MSW	Municipal Solid Waste
NCP	National Contingency Plan
NEPA	National Environmental Policy Act
NPDES	National Pollutant Discharge Elimination System
NPL	National Priority List
NRC	National Response Center
OSHA	Occupational Safety and Health Administration
PCBs	Polychlorinated Biphenyls
POTW	Publicly Owned Treatment Works
PPA	Pollution Prevention Act of 1990
PPIC	Pollution Prevention Information Clearinghouse
PRP	Potentially Responsible Party
RCRA	Resource Conservation and Recovery Act
RI/FS	Remedial Investigation / Feasability Study
RQ	Reportable Quantity
SARA	Superfund Amendments and Reauthorization Act
SDWA	Safe Drinking Water Act
SERC	State Emergency Response Commission
SIP	State Implementation Plan

SPCC	Spill Prevention, Control and Countermeasures
SPQ	Small Quantity Generator
TPQ	Threshold Planning Quantity
TRI	Toxic Release Inventory
TSCA	Toxic Substances Control Act
TSDF	Treatment, Storage, or Disposal Facility
UST	Underground Storage Tank
WQS	Water Quality Standards

PART I INTRODUCTION TO ENVIRONMENTAL LAW

Chapter 1

Federal Environmental Law - Overview

1.1   INTRODUCTION

The purpose of this chapter is to provide an overview of the
different federal environmental laws that impact various industrial and
business operations. With the steady poliferation of pollution control
laws and regulations, it is important for every company to understand
how these laws may affect a company's daily business operations. Each
federal environmental law seeks to achieve its own pollution-related
goals concerning different types of pollutants that affect different
types of environmental media, i.e., ground, water, and air. For example,
the federal Resource Conservation and Recovery Act (RCRA) sets forth a
"cradle-to-grave" regulatory framework for management of hazardous and

solid wastes;[1] the Clean Air Act places controls on the emission of air pollutants into the ambient air;[2] the Clean Water Act seeks to protect the physical integrity of U.S. waters;[3] the Comprehensive Environmental Response Compensation and Liability Act (CERCLA) governs the clean up of hazardous substances that pose a threat to the environment,[4] and the Emergency Planning and Community Right-to-Know Act (EPCRA) facilitates emergency response planning for chemical hazards.[5] The purpose and regulatory framework of each of these laws are summarized here in order to provide the reader with a working knowledge of federal environmental laws that may affect industrial and business operations. Obviously, various state environmental law counterparts, which in large part parallel the federal regime, may also impact a company's environmental management decisions. Chapter 2 follows with a summary of state environmental law requirements.

1.2 COMPREHENSIVE ENVIRONMENTAL RESPONSE, COMPENSATION AND LIABILITY ACT (CERCLA)

In 1980, Congress enacted the Comprehensive Environmental Response, Compensation and Liability Act (CERCLA),[6] commonly known as "Superfund," to facilitate remedial action whenever there is a release - or threatened release - of hazardous substances into the environment, and to hold responsible parties liable for any cleanup and response costs. A litigation explosion has occurred since CERCLA became law in 1980. Due to the enormous costs associated with the cleanup of contaminated Superfund sites, potentially responsible parties (PRPs) have become pitted against each other, their insurance carriers, and the government in court while attempting to minimize or escape the wrath of hazardous waste liability. With the average cost to clean up a site running in the neighborhood of thirty million dollars, CERCLA liability can be financially devastating to potentially responsible parties.

Liability is established under CERCLA Section 107(a) if:[7]

> 1. The contaminated site in question is a "facility" as defined in CERCLA Section 101(9);[8]
> 2. The defendant is a responsible person under CERCLA Section 107(a);[9]

---

1 See 1.4.

2 See 1.7.

3 See 1.6.

4 See 1.2.

5 See 1.5.

6 42 USC 9601 et seq.

7 Town of Munster v. Sherwin-Williams Co., 27 F.3d 1268 (7th Cir. 1994); Kerr-McGee Chemical v. Lefton Iron & Metal, 14 F.2d 321 (7th Cir. 1994).

8 42 USC 9601(9). See Elf Atochem North America, Inc. v. United States, 1994 WL 559219 (E.D. Pa. Sept. 22, 1994).

9 See 1.2.2.

3. A release or a threatened release[10] of a hazardous substance[11] has occurred; and
4. The release or the threatened release has caused the government or a private party to incur response costs.[12]

1.2.1 Strict Liability; Joint and Several Liability

CERCLA Section 107(a)[13] imposes "strict liability" on potentially responsible parties (PRPs) for response costs resulting from a release of hazardous substances. In other words, it is unnecessary for the government or a private party to prove that the owner or operator of a facility was negligent or otherwise caused the release.[14] It is merely necessary to establish that a hazardous substance was "released" at the site.[15]

Thus, in an action under CERCLA for recovery of cleanup or response costs, the government or private party plaintiff need only prove a "nexus" between the defendant and the hazardous substances released at the site.[16] For example, one federal district court has stated that "CERCLA section 107 requires only a minimal causal nexus between the defendant's hazardous waste and the harm caused.... CERCLA only requires that the plaintiff prove by a preponderance of the evidence that the defendant deposited his hazardous waste at the site and that the hazardous substances containing the defendant's waste are also found at the site."[17] There is no need to a prove the traditional elements of causation when asserting a claim under CERCLA.[18]

---

10 See Companies for Fair Allocation v. Axil Corp., 853 F. Supp. 575 (D. Conn. 1994); Bunger v. Hartman, 851 F. Supp. 461 (S.D. Fla. 1994).

11 42 USC 9601(14). The term "hazardous substance" is defined broadly in CERCLA by reference to substances defined as hazardous in a number of other environmental statutes, including the Resource Conservation and Recovery Act, 42 USC 6921; the Clean Water Act, 33 USC 1317(a); and the Clean Air Act, 42 USC 7412.

12 See generally Reitze, Harrison & Palko, "Cost Recovery by Private Parties under CERCLA: Planning a Response Action for Maximum Recovery," 27 Tulsa L.J. 365 (Spring 1992).

13 42 USC 9607(a).

14 See, e.g., Kerr-McGee Chemical Corp. v. Lefton Iron & Metal Co., 14 F.3d 321, 325-26 (7th Cir. 1994); In re Chicago, Milwaukee, St. Paul & Pacific R.R., 974 F.2d 775, 779 (7th Cir. 1992).

15 See Environmental Transportation Systems v. Ensco, Inc., 969 F.2d 503 (7th Cir.1992); New York v. Shore Realty Corp., 759 F.2d 1032 (2d Cir.1985); United States v. Conservation Chem. Co., 619 F. Supp. 162 (W.D. Mo. 1985); Mardan Corp. v. C.G.C. Music Ltd., 600 F. Supp. 1049 (D. Ariz. 1984), aff'd, 804 F.2d 1454 (9th Cir. 1986); United States v. Northeastern Pharmaceutical & Chem. Co. ("NEPACCO"), 579 F. Supp. 823 (W.D. Mo. 1984), aff'd in part, rev'd in part, 810 F.2d 726 (8th Cir. 1986) cert. denied 484 U.S. 848 (1987).

16 See United States v. Maryland Sand, Gravel and Stone Co., 1994 WL 541069 (D. Md. Aug. 12, 1994).

17 Violet v. Picillo, 648 F. Supp. 1283 (D. R.I. 1986). See also Acme Printing Ink Co. v. Menard, Inc., 1994 WL 692963 (E.D. Wis. Dec. 5, 1994).

18 See Northwestern Mutual Life Ins. Co. v. Atlantic Research Corp., 847 F. Supp. 389 (E.D. Va. 1994); United States v. Stringfellow, 661 F. Supp. 1053 (C.D. Cal. 1987); United States v. Bliss, 667 F. Supp. 1298 (E.D. Mo. 1987).

CERCLA also imposes "joint and several liability" on PRPs.[19] When liability is joint and several, any single defendant can be responsible for the entire cost of a cleanup or other response costs incurred at a site. Joint and several liability is traditionally imposed when the action of two or more defendants causes a single indivisible result.[20]

1.2.2 Potentially Responsible Parties

CERCLA casts a wide liability net to capture several classes of PRPs. Congress designated four broad categories of PRPs who, regardless of fault, are liable for Superfund cleanups if they contributed any amount of a hazardous substance to the site.[21] The four classes of PRPs are:

1. Current owners or operators of a site;[22]

2. Past owners or operators of a site at the time hazardous substances were disposed of at the site;[23]

3. Anyone who arranged for the disposal, transport or treatment of hazardous substances found at the site (generators or "arrangers");[24] and

4. Anyone who accepted hazardous substances for disposal and selected the site now slated for cleanup (transporters).[25]

Judicial interpretation of Section 107 has expanded this list of PRPs to include successor corporations to generators, transporters or former owners or operators;[26] lessees of current and former landowners;[27]

---

19 See New York v. Shore Realty Corp., 759 F.2d 1032 (2d Cir. 1985); United States v. New Castle County, 642 F. Supp. 1258 (D. Del. 1986); United States v. Medley, 24 Env't Rep. Cas. (BNA) 1856 (D.S.C. 1986); United States v. Conservation Chem. Co., 589 F. Supp. 59 (W.D. Mo. 1984); United States v. A & F Materials Co., Inc., 578 F. Supp. 1249 (S.D. Ill. 1984).

20 See United States v. DWC Trust Holding Co., 1994 WL 395730 (D. Md. July 22, 1994).

21 CERCLA 107(a); 42 USC 9607(a).

22 42 USC 9607(a)(1). See Kerr-McGee Chemical Corp. v. Lefton Iron & Metal Co., 14 F.3d 321 (7th Cir. 1994); G.J. Leasing Co. v. Union Electric Co., 854 F. Supp. 539 (S.D. Ill. 1994).

23 42 USC 9607(a)(2). See Joselyn Manufacturing Co. v. Koppers Co., Inc., 40 F.3d 750 (5th Cir. 1994).

24 42 USC 9607(a)(3). See Saint Paul Fire & Marine Ins. Co. v. Warwick Dyeing Corp., 26 F.3d 1195 (1st Cir. 1994); City of Detroit v. A.W. Miller, Inc., 842 F. Supp. 957 (E.D. Mich. 1994).

25 42 USC 9607(a)(4). See Tippins, Inc. v. USX Corp., 37 F.3d 87 (3rd Cir. 1994); Atlantic Richfield Co. v. Blosenski, 847 F. Supp. 1261 (E.D. Pa. 1994).

26 See Kleen Laundry & Dry Cleaning Services v. Total Waste Management, 867 F. Supp. 1136 (D.N.H. 1994).

27 See Northwestern Mutual Life Ins. Co. v. Atlantic Research Corp., 847 F. Supp. 389 (E.D. Va. 1994); Folino v. Hampden Color and Chemical Co., 1993 WL 359845 (D. Vt. Sept. 2, 1993); Caldwell v. Gurley Refining Co., 755 F.2d 645 (8th Cir. 1985); United States v. Argent Corp., 21 Env't Rep. Cas. (BNA) 1354 (D. N.M. 1984).

corporate officers who were active in site operations;[28] active shareholders;[29] parent corporations;[30] lenders;[31] and trustees.[32]

Once a site is targeted for cleanup, EPA has several remedial options to pursue. Generally, EPA will send a "PRP letter" announcing that the recipient has been designated as a PRP for a particular site. The EPA then undertakes settlement negotiations with the PRP for either cleanup costs incurred by the government or to induce the PRP to commence voluntary cleanup of the site. If settlement negotiations fail with one or all of the PRPs, Superfund provides EPA with two mechanisms to force cleanup of the site. First, CERCLA Section 104 authorizes the EPA to remove life-threatening toxic materials,[33] and then to sue PRPs under Section 107 for funds spent on cleanup.[34] The second alternative is for EPA to order an abatement action under CERCLA Section 106.[35]

The typical Superfund site involves generation and disposal of hazardous waste over several decades by numerous companies, many of which may be insolvent, judgment proof, or otherwise no longer in existence at the time of cleanup. Although a single party is rarely responsible for the entire amount of hazardous substances disposed at a site, EPA is not required to sue all PRPs. EPA can file suit against a single PRP who was a relatively minor contributor of hazardous substances found at the site and seek response costs for cleanup of the entire site from that one PRP. Moreover, EPA is under no duty to investigate whether there are any other PRPs or to inform the chosen PRP about the existence of other PRPs. Thus, the targeted PRP must bear the cost of investigating the site for any other PRPs and the cost of litigating for contribution against other PRPs, even if the targeted PRP is a relatively minor contributor of wastes to the site.

---

28 See Sydney S. Arst Co. v. Pipefitters Welfare Education Fund, 25 F.3d 417 (7th Cir. 1994); United States v. Northeastern Pharmaceutical & Chem. Co. ("NEPACCO"), 810 F.2d 726 (8th Cir. 1986), cert. denied, 108 S.Ct. 146 (1987); United States v. Ward, 618 F. Supp. 884 (E.D.N.C. 1985).

29 See Truck Components, Inc. v. Beatrice Co., 1994 WL 520939 (N.D. Ill. Sept. 21, 1994); Jacksonville Electric Auth. v. Bernuth Corp. 996 F.2d 1107 (11th Cir. 1993); John S. Boyd Co. v. Boston Gas Co., 992 F.2d 401 (1st Cir. 1993).

30 See United States v. TIC Investment Corp., 866 F. Supp. 1173 (N.D. Iowa 1994); Bronson Specialties, Inc. v. Houghton, 1993 U.S. Dist. LEXIS 7714 (W.D. Mich. Apr. 12, 1993); Kelley v. Thomas Solvent Co., 727 F. Supp. 1554 (W.D. Mich. 1989). But see Joslyn Corp. v. T.L. James Co., Inc., 696 F. Supp. 222 (W.D. La. 1988), aff'd, 893 F.2d 80 (5th Cir. 1990).

31 See United States v. Fleet Factors Corp. (Fleet Factors IV), 821 F. Supp. 707 (S.D. Ga. 1993); United States v. Maryland National Bank & Trust Co., 632 F. Supp. 573 (D. Md. 1986); United States v. Mirabile, 18 Envt'l L. Rep. (ELI) 20994 (E.D. Pa. 1985). See generally Dennison, Keim, and Lee, Environmental Due Diligence for Lenders (Warren, Gorham & Lamont, 1995).

32 See North Carolina v. W.R. Peele, Sr. Trust, 1994 U.S. Dist. LEXIS 16335 (E.D.N.C. 1994); City of Phoenix v. Garbage Services Co., 816 F. Supp. 564 (D. Ariz. 1993).

33 42 USC 9604.

34 42 USC 9607.

35 42 USC 9606.

1.2.3 Defenses to CERCLA Liability

The CERCLA statute sets forth a limited number of defenses to liability.[36] PRPs may escape liability through successful assertion of one of the following enumerated defenses:

1. Act of God;[37]
2. Act of war;[38]
3. Act or omission of a third party;[39]
4. Innocent landowner defense;[40] and
5. Security interest exemption.[41]

The act of God and act of war defenses have rarely been used.[42] To avoid liability, PRPs have primarily relied on the third-party defense and innocent landowner defense, neither of which has been particularly successful. There is conflicting opinion among the courts on whether equitable defenses may also be available. EPA takes the position that the only defenses available are those listed in the statute.[43] Some courts have, however, recognized equitable defenses to CERCLA liability.[44]

Third-Party Defense

The third-party defense was envisioned to protect owners and operators from the so-called "midnight dumper," a third party who dumps hazardous substances on property without knowledge of the property owner. This defense to liability is only available when a release occurred solely because of "an act or omission of a third party" and the PRP asserting the defense exercised due care with respect to the hazardous substance and took precautionary measures against foreseeable acts or omissions of third parties.[45] The defense is not available if the release is caused by (1) an act or omission of an agent of the

---

36 CERCLA 107(b); 42 USC 9607(b).

37 CERCLA 107(b)(1); 42 USC 9607(b)(1).

38 CERCLA 107(b)(2); 42 USC 9607(b)(2).

39 CERCLA 107(b)(3); 42 USC 9607(b)(3).

40 CERCLA 101(35)(A); 42 USC 9601(35)(A).

41 CERCLA 101(20)(A); 42 USC 9601(20)(A),

42 But see Wagner Seed Co. v. Daggett, 800 F.2d 310 (2d Cir. 1986) (act of God defense); United States v. Shell Oil Co., 34 Env't Rep. Cas. (BNA) 1342 (C.D. Cal. 1992) (act of war defense).

43 See United States v. Smuggler-Durant Mining Corp., 823 F. Supp. 873 (D. Colo. 1993); United States v. Hardage, 26 Env't Rep. Cas. (BNA) 1049 (W.D. Okla. 1987).

44 See Thaler v. PRB Metal Products, Inc., 815 F. Supp. 99 (E.D.N.Y. 1993); Allied Corp. v. Acme Solvents Reclaiming, Inc., 691 F. Supp. 1100 (N.D. Ill. 1988); General Electric Co. v. Litton Business Systems, Inc. 715 F. Supp. 949 (W.D. Mo. 1989).

45 CERCLA 107(b)(3); 42 USC 9607(b)(3).

defendant or (2) an act or omission of a third party with whom the defendant has a contractual relationship.[46]

The third-party defense is the most heavily litigated of the statutory defenses found in CERCLA Section 107(b). Under this defense, a defendant carries the burden of demonstrating that a "totally unrelated third party is the sole cause of the release."[47] A third party must be "other than one whose act or omission occurs in connection with a contractual relationship, existing directly or indirectly, with the defendant."[48] For example, several courts have made clear that a lease is a "contractual relationship," which precludes landlords and tenants from claiming third-party defenses with regard to each other's actions.[49] Similarly, courts have found that a generator's business relationship with a waste transporter prevents successful assertion of the third-party defense even when the generator was unaware of the location to which the transporter was taking its wastes,[50] as well as when it arranged to have the wastes disposed at one site and the transporter diverted the wastes to another site without authorization.[51] CERCLA Section 107(b)(3) also requires the defendant to show by a preponderance of the evidence that he exercised "due care with respect to the hazardous substance concerned," and that he "took precautions against foreseeable acts or omissions" of any such third party.[52]

Innocent Landowner Defense

The "innocent landowner" defense was added to CERCLA in 1986 with passage of the Superfund Amendments and Reauthorization Act (SARA).[53] The defense was expected to moderate Superfund liability by excluding from the group of potentially responsible parties those "innocent" landowners who:

  1. Did not know that the property was contaminated at the time of purchase;

46 See United States v. Maryland Sand, Gravel and Stone Co., 1994 WL 541069 (D. Md. Aug. 12, 1994); Chatham Steel Corp. v. Brown, 858 F. Supp. 1130 (N.D. Fla. 1994).
47 See G.J. Leasing Co. v. Union Electric Co., 854 F. Supp. 539 (S.D. Ill. 1994); New York v. Laskins Arcade Co., 856 F. Supp. 153 (S.D.N.Y. 1994); O'Neil v. Picillo, 682 F. Supp. 706 (D.R.I. 1988), aff'd, 883 F.2d 176 (11th Cir. 1989), cert. denied, 110 S. Ct. 1115 (1990).
48 CERCLA 107(b)(3), 42 USC 9607(b)(3).
49 See United States v. Monsanto, 858 F.2d 160 (4th Cir. 1988), cert. denied, 490 U.S. 1106 (1989); International Clinical Laboratories, Inc. v. Stevens, 710 F. Supp. 466 (E.D.N.Y. 1989); United States v. Northernaire Plating Co., 670 F. Supp. 742 (W.D. Mich. 1987).
50 See United States v. Mottolo, 695 F. Supp. 615 (D.N.H. 1988). See also United States v. Parsons, 723 F. Supp. 757 (N.D.Ga. 1989).
51 See O'Neil v. Picillo, 682 F. Supp. 706 (D.R.I. 1988), aff'd, 883 F.2d 176 (1st Cir. 1989), cert. denied, 110 S. Ct. 1115 (1990).
52 CERCLA 107(b)(3)(a), (b); 42 USC 9607(b)(3)(a), (b). Section 107(b)(3) also states that "an employee or agent of the defendant" is not a third party for the purposes of the defense.
53 CERCLA 101(35)(A); 42 USC 9601(35)(A).

2. Reacted responsibly to the contamination when found; and

3. Made reasonable inquiries into the past uses of the property before purchase to determine whether the property was contaminated.

The innocent landowner defense to CERCLA liability provides that a potentially responsible party (PRP) may avoid liability by establishing that property was acquired by the PRP after the disposal or placement of hazardous substances on the property and that the PRP did not know, nor have any reason to know, about the presence of any hazardous substances.[54] In addition, the new property owner must, at the time of purchase, make appropriate inquiry into the previous ownership and uses of the property and take steps to minimize liability consistent with good commercial or customary practice.[55] Courts have, however, generally been reluctant to find that a PRP qualifies for the innocent purchaser defense.[56] Paradoxically, however, a few courts have allowed the innocent landowner defense in cases where the purchaser failed to inspect the property prior to its acquisition.[57]

Security Interest Exemption

CERCLA Section 101(20)(A) limits the scope of owner/operator liability by exempting from liability a person "who, without participating in the management of a vessel or facility, holds indicia of ownership primarily to protect his security interest."[58] This provision has generally been construed as shielding lenders from liability that might otherwise result from holding a security interest in contaminated property, provided they have not participated in the management of the day-to-day activities of the facility.[59] Even with the security interest exemption, lenders remain at risk for cleanup liability. The limited exemption has provided little assurance that a lender will not find itself party to a costly CERCLA litigation. If the lender holds a mortgage on property and becomes involved in the operation or management of the property, it may be found liable as an

---

54 See Cross, "Establishing Environmental Innocence," 23 Real Estate L.J. 332 (Spring 1995).

55 See In re Hemingway Transport, Inc., 993 F.2d 915 (1st Cir. 1993); United States v. A & N Cleaners & Launderers, Inc., 854 F. Supp. 229 (S.D.N.Y. 1994).

56 See Kerr-McGee Chemical Corp. v. Lefton Iron & Metal Co., 14 F.3d 321 (7th Cir. 1994); United States v. Shell Oil Co., 841 F. Supp. 962 (C.D. Cal. 1993); United States v. Broderick Investment Co., 862 F. Supp. 272 (D. Colo. 1994); Acme Printing Ink Co. v. Menard, Inc., 1994 WL 692963 (E.D. Wis. Dec. 5 , 1994); Wickland Oil Terminals v. ASARCO Inc., 18 ELR 20855 (N.D. Cal. 1988). But see International Clinical Laboratories, Inc. v. Stevens, 710 F. Supp. 466 (E.D.N.Y. 1989).

57 See U.S. v. Pacific Hide & Fur Depot, Inc., 716 F. Supp. 1341 (D. Idaho 1989); United States v. Serafini, 706 F. Supp. 346 (M.D. Pa. 1988).

58 42 USC 9601(20)(A).

59 The language of the security interest exemption states clearly that the lender must not participate in the management of the property in which he holds the security interest to be entitled to the exemption. 42 USC 9601(20)(A), CERCLA 101(20)(A).

"operator of the facility."[60] Likewise, if the lender forecloses on the property and acquires the property at a foreclosure sale, it may become liable as an "owner of the facility."[61]

In April 1992, EPA issued regulations which provided a framework of specific tests designed to clarify the scope of lender liability under CERCLA.[62] The EPA rule provided an overall standard for judging when a lender's participation in management would cause the lender to forfeit its exemption.[63] According to these regulations, a lender could - without incurring liability - undertake investigatory actions before the creation of a security interest, monitor or inspect the property, and require that the borrower comply with all environmental standards.[64] When a loan neared default, the rule allowed the lender to engage in work-out negotiations and activities, including ensuring that the collateral property did not violate environmental laws.[65] The rule also was designed to protect a secured creditor that acquires full title to the collateral property through foreclosure, as long as the creditor did not participate in the property's management prior to foreclosure and made certain diligent efforts to divest itself of the property.[66]

Significantly, the Chemical Manufacturers Association brought suit to contest the validity of the EPA regulations, hoping to ensure the availability of a larger pool of PRPs in the event that a judgment was rendered against any of its members. In 1995, the U.S. Court of Appeals for the D.C. Circuit invalidated the EPA's lender liability rules, holding that the EPA had exceeded its rulemaking authority by adopting regulations designed to affect the ability of private parties to maintain an action under CERCLA Section 107.[67] The court stated that "Congress, by providing for private rights of action under section 107, has designated the courts and not EPA as the adjudicator of the scope of CERCLA liability."[68]

Although the DC Circuit court has invalidated the EPA regulations, the EPA will likely continue to use these rules as agency guidance concerning a lender's potential liability to the EPA. Thus, even though these rules are no longer legally enforceable as regulations, lenders may consider them as agency guidelines regarding suits brought by the EPA. Moreover, lenders may still be protected by the security interest exemption. They simply may no longer look to the EPA regulations as proof of entitlement to the defense. Rather, courts will determine on a

---

60 42 USC 9607(a)(2).

61 42 USC 9607(a)(1), CERCLA 107(a)(1).]

62 40 CFR 300.1100.

63 40 CFR 300.1100(c)(1) (1992).

64 40 CFR 300.1100(c)(2).

65 40 CFR 300.1100(c)(2)(ii)(B).

66 40 CFR 300.1100(d).

67 Kelley v. EPA, 1994 WL 27881 (D.C.Cir. Feb. 4, 1994).

68 Kelley v. EPA, 1994 WL 27881 at *6.

case-by-case basis whether a lender is shielded from liability by virtue of the security interest exemption of CERCLA Section 101(20)(A).

1.2.4 Corporate Liability

Where a subsidiary corporation is deemed a PRP by EPA, the parent corporation is also likely to be sued as a defendant under the legal theory known as "piercing the corporate veil." The theory is generally advanced on the basis that the parent corporation controls the actions of its subsidiary. EPA may rely on the following types of evidence to pierce the corporate veil and add the parent corporation as a PRP:

* Representation of the parent on the board of directors of the subsidiary;
* Individuals acting as officers for the parent and the subsidiary corporation;
* Oversight by the parent of the subsidiary's finances and operating procedures.

The determinative factor of whether the corporate veil can be pierced is the degree of control the parent exercises over the subsidiary's affairs. Many courts have found sufficient grounds to impose CERCLA liability on a parent corporation.[69] However, in cases where the parent and subsidiary faithfully adhere to basic corporate formalities, such as keeping separate books and records, making separate arrangements for employee benefits, and filing separate tax returns, the parent corporation would be less likely to incur liability. The more autonomy the subsidiary is given the less likely that the parent corporation will be held accountable for its subsidiary's environmental liabilities.[70]

Under CERCLA, corporate officers and directors may be held personally accountable for environmental transgressions of the corporation. In determining whether personal liability is justified, courts generally consider such factors as:

* The officer's stock ownership in the corporation;
* The officer's active participation in the management of the corporation; and
* The officer's authority to control the corporation's waste handling practices.

---

69 See Atlantic Richfield Co. v. Blosenski, 847 F. Supp. 1261 (E.D. Pa. 1994); U.S. v. Kayser-Roth Corp., 724 F. Supp. 15 (D.R.I. 1989), aff'd, 910 F.2d 24 (1st Cir. 1990), cert. denied, 111 S.Ct. 957 (U.S. 1991); Colorado v. Idarado Mining Company, 916 F.2d 1486 (10th Cir. 1990); Mobay Corporation v. Allied-Signal, Inc., 761 F.Supp. 345 (D.N.J. 1991).

70 See Josyln Corp. v. T.L. James & Co., Inc., 696 F. Supp. 222 (W.D. La. 1988), aff'd, 893 F.2d 80 (5th Cir. 1990).

These same factors may be used to hold a corporate shareholder accountable for the corporation's hazardous waste liabilities.[71]

In one illustrative case, a federal district court ruled on summary judgment motions concerning the liability of various parties for costs incurred by the city of North Miami to clean up hazardous substances at a landfill site.[72] The court evaluated the degree of control, either possessed or actually exercised, that each party had over activities at the site to find two principal shareholders and officers of a land development and landfill operating company liable as "operators" under CERCLA Section 107(a), while dismissing claims of CERCLA liability against the corporate secretary, the corporate attorney, and an independent contractor.

Another case tells the story of a corporate officer's liability for hazardous waste cleanup costs. In a CERCLA private cost recovery action, the Chief Executive Officer of a company that had purchased contaminated property was found liable to the tune of 25% of the costs of a private party's cleanup action, or $311,368.03.[73]

The court concluded that the CEO was individually liable under CERCLA as an "operator" because he had exercised actual control over the corporations that owned the site and had personally participated in decisions relating to the hazardous condition at the site. The court cited a litany of actions taken by the CEO that made him subject to "operator" liability:

> "He made the decision for [the corporation] to purchase the property. He searched for potential purchasers of the hazardous material. He authorized the cleanup and closure agreements made to state and federal officials. On behalf of [the corporation], he signed the agreement with plaintiffs that they would jointly comply with the EPA's unilateral administrative order and conduct removal activities on the site. He made the decision to reopen operations at the site after deciding to set up [the sister corporation]. He was chairman of the board of [that corporation], and he controlled its finances. The president of [the sister corporation] sent [him] regular reports concerning the

---

71 Cases concerning the personal liability of corporate officers and shareholders include Jacksonville Electric Auth. v. Bernuth Corp. 996 F.2d 1107 (11th Cir. 1993); Lansford-Coaldale Joint Water Auth. v. Tonolli Corp, 4 F.3d 1209 (3d Cir 1993); Truck Components, Inc. v. Beatrice Co., 1994 WL 520939 (N.D. Ill. Sept. 21, 1994); Riverside Market Development Corp. v. International Building Products, Inc., 931 F.2d 327 (5th Cir. 1991); Kelly ex rel. Michigan Natural Resources Commission v. Arco Industries, 723 F. Supp. 1214 (W.D. Mich. 1989); U.S. v. Conservation Chemical Co., 628 F. Supp. 270 (N.D. Ill. 1990); Quadion Corporation v. Mache, 738 F. Supp. 270 (N.D. Ill. 1990); Wisconsin v. RollFink, 475 N.W.2d 575 (Wis. 1991).

72 City of North Miami v. Berger, 828 F. Supp. 401 (E.D. Va. 1993).

73 FMC Corp. v. Aero Industries, Inc., 998 F.2d 2079 (10th Cir. 1993).

operation and handling of hazardous wastes. When it proved
too difficult to put the site back into full operation, [he]
attempted to merge [the sister corporation] with another
company. When the merger failed, he decided to stop [the
company's] business operations and abandoned the site. He
negotiated the sale of the equipment on the site and
arranged for its shipping. The evidence thus fully supports
the district court's conclusion that [he] was an
operator."[74]

Successor Liability

Another liability pit for corporations to fall into is that of
successor liability. The courts have taken a very expansive approach to
extending liability for CERCLA cleanups to successor corporations. Cases
have reflected judicial concern that successor entities might be in the
position to avoid CERCLA liability for contamination caused by
predecessor companies.[75] For example, where a successor corporation
purchased all of the shares of a predecessor corporation, it was held
liable for the environmental violations that occurred during the time
that the predecessor operated the business.[76] The mere assumption of the
customers and accounts of the predecessor enterprise has been found
sufficient to impose liability on the successor entity for the costs of
implementing remediation of the predecessor's contamination.[77] The
courts have generally agreed that CERCLA is silent on the issue of
successor liability of corporations; however, in looking to federal
common law to cure this ambiguity in the statute, successor liability is
often found.[78]

Some courts have relied on a "continuity of business enterprise"
theory to extend CERCLA liability to successor corporations.[79] For

---

74 Id.

75 See Louisiana-Pacific Corp. v. Asarco, Inc., 909 F.2d 1260 (9th Cir. 1990), declaring that
although Congress failed to address specific issue of corporate successor liability under CERCLA,
Congress did intend such liability. Likewise, dissolved corporations have not found it easy to
escape CERCLA's liability net. See Barton Solvents, Inc. v. Southwest Petro-Chem, Inc., 836 F. Supp.
757 (D. Kansas 1993) (CERCLA preempts Oklahoma law regarding dissolved corporation's capacity to be
sued); United States v. Sharon Steel Corporation, 681 F. Supp. 1492 (D. Utah 1987) (holding that
CERCLA preempts state dissolution and capacity to be sued statutes). But see Levin Metals v. Parr-
Richmond Terminal Company, 817 F.2d 1448 (9th Cir. 1987).

76 See GRM Industries v. Wickes Manufacturing Co., 749 F. Supp. 810 (W.D. Mich. 1990).

77 See Smith Land & Improvement Corp. v. Celotex Corp., 851 F.2d 86 (3rd Cir. 1988) cert.denied, 488
U.S 1029 (1989). But see City Environmental, Inc. v. U.S. Chemical Co., 814 F. Supp. 624 (E.D.
Mich. 1993); Sylvester Brothers Development Company v. Burlington Northern Railroad, 772 F.Supp. 443
(D. Minn. 1990).

78 See Atlantic Richfield Co. v.Blosenski, 847 F. Supp. 1261 (E.D. Pa. 1994); Anspec Co. v. Johnson
Controls, Inc., 922 F.2d 1240 (6th Cir. 1991).

79 See Northwestern Mutual Life Ins. Co. v. Atlantic Research Corp., 847 F. Supp. 389 (E.D. Va.
1994); Charter Township of Oshtemo v. American Cyanamid Co., 1994 U.S. Dist LEXIS 8544 (W.D. Mich.

example, in *United States v. Mexico Feed and Seed Co.*,[80] the court held that the purchasing company was liable for the seller's CERCLA obligations because it retained the same management and employees, maintained and operated the same facilities, purchased nearly all of the operating assets, continued the same business and identified itself to the public as the same company as the seller. The court in *United States v. Western Processing Company*[81] endorsed a broader version of the continuing enterprise exception. The court cited five factors to be considered under the continuing enterprise test:

1. Continuity of employees, supervisory personnel, and physical location;
2. Manufacture of the same product;
3. Retention of the name;
4. Continuity of general business operations; and
5. Purchaser holding itself out as a continuation of the seller.[82]

1.2.5 EPA Response Actions

Although the EPA has the ability to respond to "releases" and spills of hazardous substances, the EPA generally looks to PRPs to implement both immediate removal and longer-term remedial actions, except in emergency situations. In an emergency situation, the EPA can order a private party to take action regardless of whether the private party is a potentially responsible party.[83] Whenever there is a release of a "hazardous substance" or of "any pollutant or contaminant" that may present an imminent and substantial danger to the public health and welfare, the EPA is authorized to act if there is no potentially responsible party that can and will respond in a timely manner.[84] In addition to finding that the "release" or spill will "present an imminent and substantial danger to the public health and welfare," prior to using government or Superfund monies to implement a response, the EPA must, however, find that:

May 4, 1994); Blackstone Valley Electric Co. v. Stone & Webster, Inc., 867 F. Supp. 73 (D. Mass. 1994).

80 United States v. Mexico Feed and Seed Co., 764 F. Supp. 565 (E.D. Mo. 1991).

81 United States v. Western Processing Company, 751 F. Supp. 902 (W.D. Wash 1990). See also United States v. Carolina Transformer Company, 739 F. Supp. 1030 (E.D.N.C. 1989).

82 See also United States v. Distler, 741 F. Supp. 637 (W.D. Ky. 1990) (company that purchased the assets of a corporation and continued the seller corporation's business, was held liable under the expanded "continuity of business enterprise" version of this exception). But see United States v. Atlas Minerals & Chemicals, Inc., 1993 U.S. Dist. LEXIS 16578 (E.D. Pa. Nov. 22, 1993) (successor corporation not liable under "continuation of enterprise" theory); City Management Corp. v. U.S. Chemical Co., Inc., 1994 WL 621146 (6th Cir. 1994) (substantial continuation rule for successor liability held inapplicable).

83 CERCLA 106(a); 42 USC 9606(a). Failure to comply with an EPA order subjects the violator to a fine of $25,000 for each day the violation continues or for each day of non-compliance with the order. CERCLA 106(b); 42 USC 9606(b).

84 CERCLA 104(a)(1); 42 USC 9604(a)(1).

* The release is not one of a naturally occurring substance;
* The release is from products that are a part of the structure and will result in exposure within residential buildings or business or community structures;
* The release is into public or private drinking water supplies due to deterioration of the system in the ordinary use.[85]

Removal vs. Remedial Action

All response actions must be undertaken in conformity with the "National Contingency Plan (NCP)."[86] Under the NCP, response actions are categorized into two basic types: removal actions and remedial actions. A removal action is an short-term or immediate cleanup of a site.[87] Removal actions may be instituted at any hazardous waste site regardless of whether the site is listed on the National Priority List (NPL) of sites requiring cleanup. Removal actions can be entirely funded from the Superfund.[88]

A remedial action is a long-term cleanup of a site. Generally, such action is preceded by a remedial investigation and feasability study (RI/FS), which is the EPA's determination of the nature and extent of the threat, and guidance as to how to evaluate and select a remedy that is appropriate for the particular situation.[89] The remedial investigation may be conducted by the potentially responsible party or parties or by the government.[90] In most instances where a PRP has been identified, it is in that party's best interest to at least participate in, or better still, direct the remedial investigation. PRPs find that they can exercise greater control over the ultimate cleanup costs if they are involved in the RI/FS process.

The remedial investigation will determine the extent of the remedial action and, together with the feasibility study that generally follows the remedial investigation, will determine the nature of the appropriate remedy for the site. The purpose of the remedial investigation is to evaluate available information and obtain further information through sampling, monitoring and an exposure assessment to determine the necessity for, and extent of, remedial and/or removal actions. The regulations specify factors to be used in making the risk determination including population, environmental and welfare concern at risk, routes

---

85 CERCLA 104(a)(2); 42 USC 9604(a)(3).

86 See 40 CFR Part 300.

87 The law says that removal should not cost more than $2 million and not take longer than one year to complete unless EPA decides to waive these restrictions.

88 Approximately 10% of the Superfund is being spent on removal actions.

89 CERCLA 104(a)(1); 42 USC 9604(a)(1).

90 CERCLA 104(a)(1); 42 USC 9604(a)(1).

of exposure, hydrogeological factors, and the likelihood of future releases if the substance remains on the site.[91]

The feasibility study evaluates various remedial options. The regulations specify that as part of the feasibility study one alternative is to be developed in each of five categories. One of the alternatives is to the "no action" alternative. In implementing the feasibility study, an initial screening is conducted based on cost, acceptable engineering practices, and effectiveness, to eliminate unfeasible alternatives.[92] The regulations provide a list of appropriate remedial actions for various types of threats.[93]

1.3 NATIONAL ENVIRONMENTAL POLICY ACT (NEPA)

In 1969, Congress enacted the National Environmental Policy Act (NEPA) to address a growing concern over the environment and the need for environmental protection.[94] NEPA requires that all federal agencies participate in achieving environmental protection goals.[95] Section 102 sets forth procedures for federal agencies to incorporate environmental considerations in their decisionmaking processes.[96]

The most important procedural requirements are contained in NEPA's action-forcing provisions, which are designed to ensure consideration of environmental factors and public participation in major federal agency decisionmaking.[97] To satisfy NEPA's procedural requirements, agencies must prepare environmental assessments (EAs) for proposed activities,[98] unless the action falls within a categorical exclusion.[99] Based on determinations made in the EA, the agency is then required to decide whether a full environmental impact statement (EIS)[100] is needed or whether it may issue a finding of no significant impact (FONSI).[101] The federal agency may also choose to bypass the EA step and prepare an EIS for the proposed activity if it is one for which agency regulations usually require an EIS.[102]

NEPA requires that an EIS to be prepared and included "in every recommendation or report on proposals for legislation and other major federal actions significantly affecting the quality of the human

---

91 See 40 CFR 300.68(e).

92 See Guidance for Feasibility Studies under CERCLA, EPA/540/G-85/003, (June 1985).

93 See 40 CFR 300.68(j).

94 42 USC 4331.

95 42 USC 4332(2).

96 42 USC 4332(2)(C); 40 CFR 1502.3

97 42 USC 4332(2)(C).

98 40 CFR 1501.4(b), 1508.9.

99 40 CFR 1501.4(a)(2).

100 40 CFR 1501.4(c), 1508.11.

101 40 CFR 1501.4(e), 1508.13.

102 40 CFR 1501.4(a)(1).

environment...."[103] Once an agency determines that an action requires
the preparation of an EIS, it must initiate the "scoping" process,[104]
during which the agency notifies and invites all concerned agencies and
individuals to participate in the decision-making process.[105] The agency
identifies the issues to be analyzed and determines the scope and depth
of the environmental analysis. Then the agency prepares the EIS, which
concisely describes how NEPA's policies will be achieved and analyzes a
range of alternatives to the proposed action.

The environmental analysis in the EIS must take a "hard look" at the
environmental consequences of the proposed action. An agency must also
take the cumulative impact of a project into consideration when deciding
whether to prepare an EIS. Cumulative impacts can result from
individually minor but collectively significant actions taking place
over a period of time.

Most states have enacted their own "little NEPA" environmental
quality review acts modeled after the federal statute.[106] Like the
federal statute, the basic purpose of these state acts is to incorporate
consideration of environmental impacts into decisions on permit
applications and other discretionary actions taken by state agencies.

In the context of a permit application, the reviewing state agency
determines whether significant environmental impacts may result from the
proposed activity. If the state agency determines that no significant
adverse environmental impacts are associated with the permit
application, it renders a "negative declaration," and the review process
stops there. If the agency, however, finds potential environmental
impacts, it will render a "positive declaration," subjecting the permit
applicant to further more detailed review of the proposed project. A
positive declaration would then start the EIS process in motion.

1.4   RESOURCE CONSERVATION AND RECOVERY ACT (RCRA)

The Resource Conservation and Recovery Act was enacted in 1976.[107]
The statute is administered primarily by EPA. The primary goal of RCRA
is to protect water, land and air from contamination from solid wastes.
A secondary goal of the statute is to control hazardous wastes and set
standards for the proper identification, listing and handling of
hazardous wastes from generation to final disposal. EPA has established

---

103 42 USC 4332(C). See also 40 CFR 1502.3, 1508.18.

104 "[A]n early and open process for determining the scope of issues to be addressed and for
identifying the significant issues related to a proposed action." 40 CFR 1501.7, 1508.25.

105 40 CFR 1501.7(a)(1).

106 See D. Mandelker, NEPA: Law and Litigation (Clark Boardman Callaghan, 1985 & Supps).

107 42 USC 6901 et seq.

several criteria for determining whether a waste is hazardous under RCRA.[108]

RCRA Subtitle C provides the framework for hazardous waste management. The key provisions relate to listing of hazardous wastes, EPA identification number requirements for hazardous waste activities, storage requirements, and hazardous waste manifest requirement for waste shipments. Subtitle I contains the underground storage tank (UST) regulations.

RCRA authorizes EPA to set specific waste management practices for generators; transporters; and treatment, storage, and disposal (TSD) facilities. Although RCRA only requires TSD facilities to obtain permits, RCRA's recordkeeping, manifesting, and reporting requirements for generators and transporters are intended to create a framework that will account for all hazardous wastes from "cradle to grave."

In its 1984 amendments to RCRA, Congress added several new programs to complete the hazardous waste management framework. First, Congress directed EPA to promulgate restrictions on land disposal of hazardous waste (commonly known as the Land Ban). Second, Congress directed EPA to issue regulations imposing stringent groundwater monitoring requirements on all existing land disposal facilities. Third, Congress added a new corrective action program to RCRA that gives EPA the authority to order generators and TSD facilities to perform cleanups of hazardous waste releases and to undertake investigations and corrective actions at permitted facilities.

1.4.1 RCRA Underground Storage Regulation

RCRA Subtitle I, containing the UST provisions of the statute, was also enacted with the 1984 amendments. These provisions require that the EPA set national standards for the installation and operation of new USTs and the upgrading and operation of existing USTs. On September 23, 1988, the EPA promulgated regulations with technical, design, and operational standards that apply to all USTs installed after December 22, 1988.[109] In addition, Subtitle I requires the reporting of all leaks and spills from USTs and the remediation of all contamination resulting from such leaks and spills.

An underground storage tank is defined as any tank (including its connected piping) holding an "accumulation of regulated substances" that has ten percent or more of its volume underground. Federal regulations and most state regulations contain a list of tanks that are exempt from

---

108 40 CFR Part 261, Identification and Listing of Hazardous Waste, contains the elements of the hazardous waste definition.

109 40 CFR Part 280. Design standards for USTs in existence or under construction before December 22, 1988, will be phased in over a ten-year period, with release detection requirements which became effective as early as December 22, 1989, for older tanks.

regulation. Most notably, home-heating oil tanks are excluded from regulation.[110]

EPA's UST regulations, 40 C.F.R. Parts 280 and 281, apply to any person who owns or operates a UST or UST system. The term "owner" is defined in the statute generally to mean any person who owns a UST that is used for the storage, use, or dispensing of substances regulated under RCRA Subtitle I, which includes both petroleum and hazardous substances.[111] The term "operator" is very broad and means "any person in control of, or having responsibility for, the daily operation of the underground storage tank."[112]

Every owner of underground storage tanks must properly maintain these tanks, and comply with all federal and state UST regulations. The federal regulations cover the design, construction, and operation of USTs from installation to closure, require the cleanup of leaks and spills, and impose recordkeeping, reporting, and financial responsibility requirements on owners and operators of USTs.[113]

UST maintenance and regulatory compliance can be costly; however, every tank owner should keep in mind the high price of noncompliance. Owners may be liable for up to $10,000 per tank for knowingly violating the tank notification requirement, and owners and operators may be liable for up to $10,000 per tank per day for failure to comply with technical requirements.[114]

1.4.2 UST Technical Standards

The technical standards of 40 C.F.R. Part 280 include:

Subpart B - UST systems: Design, Construction, Installation, and Notification (including performance standards for new UST systems, upgrading of existing UST systems, and notification requirements);
Subpart C - General Operating Requirements (including spill and overfill control, corrosion protection, reporting and recordkeeping);
Subpart D - Release Detection; 280.50 (reporting of suspected releases);
Subpart E - Release Reporting, Investigation, and Confirmation; and
Subpart G - Out of Service UST Systems (including temporary and permanent closure).

---

110 However, some states do regulate these tanks, so it is best to check with the appropriate state environmental agency.

111 RCRA 9001(3), 42 USC 6991(3).

112 RCRA 9001(4), 42 USC 6991(4).

113 40 CFR Part 280; 42 USC 6991-6991i.

114 42 USC 6991e(d)(1) and 6991b(g).

These regulations impose obligations upon UST owners and operators, separate from the subtitle I corrective action requirements, discussed below.

1.4.3 Corrective Action Requirements

Owners and operators of UST systems containing petroleum or hazardous substances must investigate, confirm, and respond to confirmed releases.[115] These requirements include, where appropriate:

* Performing a release investigation when a release is suspected or to determine if the UST system is the source of an off-site impact;[116]
* Notifying the appropriate agencies of the release within a specified period of time;
* Taking immediate action to prevent any further release (such as removing product from the UST system);
* Containing and immediately cleaning up spills or overfills;
* Monitoring and preventing the spread of contamination into the soil and/or groundwater;
* Assembling detailed information about the site and the nature of the release;
* Removing free product to the maximum extent practicable;
* Investigating soil and groundwater contamination; and,
* In some cases, outlining and implementing a detailed corrective action plan for remediation.

1.4.4 Financial Responsibility Requirements

The financial responsibility regulations require that UST owners or operators demonstrate the ability to pay the costs of corrective action and to compensate third parties for injuries or damages resulting from the release of petroleum from USTs.[117] The regulations require all owners or operators of petroleum USTs to maintain an annual aggregate of financial assurance of $1 million or $2 million, depending on the number of USTs owned. Financial assurance options available to owners and operators include:

* Purchasing commercial environmental impairment liability insurance;
* Demonstrating self-insurance;
* Obtaining guarantees, surety bonds, or letters of credit;
* Placing the required amount into a trust fund administered by a third party; or

---

115 40 CFR 280.51 through 280.67.

116 Investigation and confirmation steps include conducting tests to determine if a leak exists in the UST or UST system and conducting a site check if tests indicate that a leak does not exist but contamination is present.

117 40 CFR Part 280, subpart H.

      * Relying on coverage provided by a state assurance fund.

1.5 EMERGENCY PLANNING AND COMMUNITY RIGHT-TO-KNOW ACT

In 1986, the Emergency Planning and Community Right-to-Know Act (EPCRA),[118] also known as SARA Title III, was added as a freestanding part of the Superfund Amendments and Reauthorization Act (SARA), which amended CERCLA. The primary goal of EPCRA is to facilitate public awareness and emergency response planning for chemical hazards. To fulfill this purpose, EPCRA requires that certain companies that manufacture, process, and use chemicals in specified quantities must file written reports, provide notification of spills/releases, and maintain toxic chemical inventories.

1.5.1 State and Local Emergency Response Planning

EPCRA provides a framework for state and local governments to design emergency response plans to handle chemical spills and releases. The governor of each state must designate a State Emergency Response Commission (SERC) to oversee the emergency planning process. Each SERC then designates Emergency Planning Districts within the state, and appoints Local Emergency Planning Committees (LEPCs) in each district. These LEPCs develop emergency response plans and establish procedures for handling public information requests. This system of state and local management provides the network for emergency response and community involvement. LEPCs must be composed of local representatives, such as elected officials, fire and police personnel, community groups, the media, and regulated companies and individuals.[119]

1.5.2 Notification to State and Local Authorities

Companies that are subject to EPCRA emergency planning requirements must notify the appropriate SERC and the LEPC within 60 days after:

> 1. Their facility acquires a listed extremely hazardous substance (EHS) in an amount above its threshold planning quantity (TPQ) or
> 2. A new substance is added to EPA's list of extremely hazardous substances and is present at the facility in an amount above the TPQ.[120]

EPA has identified over 350 EHSs. A list of these substances and their TPQs are found in Appendix A to 40 C.F.R. Part 355. Regulated companies must also designate a facility representative to notify the LEPC of any changes that occur in the manufacture, process, or use of any EHSs at

---

118 42 USC 11001-11050.

119 EPCRA 301(a)-(c); 42 USC 11001(a)-(c).

120 EPCRA 302(c), 42 USC 1002(c).

the facility.[121] Failure to give the required notice of the presence of EHSs may subject the facility to penalties of up to $25,000 per day.[122]

1.6 CLEAN WATER ACT (CWA)

The primary objective of Clean Water Act (CWA) is to "restore and maintain the chemical, physical, and biological integrity of the Nation's waters."[123] CWA Section 311 governs discharges of oil and hazardous substances into U.S. waters.

1.6.1 National Pollutant Discharge Elimination System

The current form of the CWA was enacted in 1972 and amended in 1977 and 1987.[124] Through the National Pollutant Discharge Elimination System (NPDES) permit process, the CWA regulates the point source discharges of pollutants into surface waters of the United States.[125] The U.S. EPA is delegated the authority to establish effluent limitations for industrial discharges and to require monitoring and reporting on the quantity and quality of the effluent discharged. NPDES permits are generally issued by states with EPA-approved NPDES programs. Most states have approved programs. For those states without NPDES permit authority, EPA Regional offices for those states oversee issuance of NPDES permits.

Under the NPDES system, permits are issued to different types of facilities that discharge pollutants into U.S. waters. A NPDES permit is essentially a license that allows a facility to discharge contaminated water. On the NPDES permit application, the facility provides information about the type of facility and type of discharge being requested. If the facility's application is approved, the permitting authority will issue a permit that contains various conditions related to the facility's pollutant discharges. The permit will generally contain specific limitations on contamination levels or specific actions that the facility must take, such as sampling or inspections.

1.6.2 Issuance of NPDES Individual Permits

There are two different types of permits that are issued under the Clean Water Act's NPDES program: individual permits and general permits. An individual permit is issued for one particular facility. A permit writer reviews the facility's NPDES application, may make a site visit to better understand the facility's operation, and then tailors the individual permit to address specific conditions at the facility. Although the permit is individualized, a large portion of the permit

---

121 EPCRA 303(d)(2), 42 USC 11003(d)(2).

122 EPCRA 325(a), 42 USC 11045(a).

123 CWA 101, 33 USC 1251(a).

124 The 1995 reauthorization and amendment of the CWA was pending at the time of publication.

125 Under 402(p) of the Clean Water Act, the NPDES permit program also addresses stormwater discharges.

will, however, consist of standard conditions. These conditions are included in all permits issued by the permitting authority. Standard conditions often include the following:

* Duty to provide information
* Signatory requirements
* Inspection and entry
* Penalties for falsification of records
* Definitions
* Notification requirements
* Nontranferability
* Use of proper operation and maintenance
* Halt production for noncompliance
* Duty to reapply

Facility-specific conditions usually include monitoring requirements which require that the facility take samples of the discharges into U.S. waters and analyze the contaminant levels of specified pollutants. The most commonly requested pollutants to be analyzed are pH, total suspended solids, and oil and grease. Sampling of additional pollutants may be required if they are suspected to be in the facility's discharges.

1.7 CLEAN AIR ACT (CAA)

The Clean Air Act gives EPA authority to set national ambient air quality standards for protecting public health and the environment from pollutants in the outdoor air. Primary standards set limits to protect public health, including the health of people particularly sensitive to air pollution, such as children, the elderly,and those with respiratory problems. Secondary standards set limits to protect plants, wildlife, building materials, and historic monuments.

While EPA sets standards and national regulations for controlling air pollution, it is the states that manage most of the specific programs for achieving these standards. State implementation plans (SIPs) are legally enforceable documents that state governments develop to identify their sources of air pollution, and to determine what reductions they must make to meet the federal air quality standards. Based on these determinations, measures are developed to achieve the necessary air pollution reductions.

1.7.1 Clean Air Act Amendments of 1990

On November 15, 1990, important changes to the Clean Air Act were signed into law with the Clean Air Act Amendments of 1990. The 1990 Amendments place new federal controls on small sources of air pollution that may ultimately affect hundreds of thousands of companies. In addition, Section 507 of the Clean Air Act Amendments requires that all state governments establish Small Business Technical and Environmental

Compliance Assistance Programs to help small businesses contend with various air pollution control requirements. Small business owners should ask the state pollution control agency for information about the assistance program in their state.

The Clean Air Act Amendments of 1990 contain several new requirements that are of special concern for businesses. These include measures to:

* Lower emissions from small industrial and service companies that contribute to ground level ozone pollution (smog);
* Reduce automotive emissions by establishing tailpipe inspection and maintenance programs for motor vehicles, and by expediting the development of clean automotive fuels and new motor vehicles that emit very little pollution;
* Sharply curb emissions of 189 toxic air pollutants from hundreds of industries;
* Prevent or minimize the risks from the accidental release of 100 or more very hazardous chemicals into the air;
* Recycle and phase out the production and use of products and substances that deplete the Earth's upper ozone layer; and
* Require many sources affected by the Clean Air Act to document their air pollution control obligations in a five-year operating permit.

Under the Clean Air Act Amendments of 1990, businesses will be affected by controls of three types of air pollution:

1. Primary or "Criteria" Pollutants;
2. Hazardous Air Pollutants; and
3. Ozone Depleters.

1.7.2 Primary Pollutants

EPA has already set standards for six primary (so-called "criteria") pollutants which are generally discharged in large quantities by a wide variety of sources in urban and other areas of the country. The six pollutants are ground level ozone or "smog," carbon monoxide (CO), particulate matter, nitrogen dioxide (NO_2), sulfur dioxide (SO_2), and lead. None are thought to be carcinogenic, but exposure to high or even moderate levels for varying periods of time contributes to respiratory diseases, heart ailments, and blood and circulatory problems. Control measures for ground level ozone (smog) are expected to have particularly significant effects on many small businesses.

1.7.3 Hazardous Air Pollutants

Toxic or hazardous air pollutants include chemicals that are known to cause, or that are suspected to cause, cancer and other serious health effects. The 1990 Amendments distinguish between toxic air pollutants that enter the air from routine emissions, and hazardous substances that are especially dangerous when accidentally released into the air.

EPA is responsible for regulating the routine (generally daily) emissions of toxic air pollutants under the National Emissions Standards for Hazardous Air Pollutants (NESHAP) program. About half of all air toxics emissions come from cars and other mobile sources while the other half is emitted by large and small stationary sources. The 1990 Amendments govern both of these types of sources, but it is the stationary source controls that have the most direct impact on small businesses. The Act requires EPA to set toxic air pollutant standards for specific industry activities. EPA has identified several of these, including solvent degreasing, surface coating processes, agricultural chemical production, paint stripping operations, dry cleaning, wood treatment, and many other industrial activities.

The Act requires that EPA establish emissions standards for "categories" of affected sources. The standards apply for all "major sources" and, in some cases, for smaller so-called "area sources." A small business can be either a "major source," an "area source," or be completely unaffected by the air toxic requirements. The distinction relates only to how much of one or more toxic pollutants a business emits into the air. The actual size of the business, the volume of goods or services it produces, and the number of employees do not necessarily correlate with how much air pollution is generated. Small businesses that may be affected by the air toxics provisions include gasoline stations, auto body repair shops, metal finishers, surface coating and painting operations, and solvent degreasing operations.

Source Categories

Any source of toxic air pollution that emits 10 tons per year of any listed toxic air pollutant, or combination of 25 tons or more, will be regulated as a major source of toxic air pollution. Over the next ten years, these major sources will be required to install the best proven air pollution control technology for their particular industry.

The Act also gives EPA the discretion to regulate certain other sources as "major sources" even though they emit less than the 10/25 ton limit. Lesser quantities can be set for pollutants that are highly toxic to human health and the environment.

EPA will establish and phase in performance standards for each industry (source category) based on what is termed "Maximum Achievable Control Technology (MACT)." All "major sources" will be subject to these MACT controls, which are designed to reduce emissions of air toxics significantly over the next decade. The Act allows any source to obtain a six-year extension from full compliance with the MACT if it reduces air toxics emissions by 90 to 95 percent before the applicable MACT standard is proposed. This may be a strong incentive for small businesses because it is often easier and cheaper to reduce the bulk of a source's excess emissions than it is that last fraction needed to achieve a specific limitation. For small businesses that need to reduce any of the listed pollutants, early reduction could avoid hours of paperwork and other administrative work.

Many small businesses will also be affected by controls on "area sources" of toxic air pollutants. These smaller sources emit less than 10 tons per year of a single air toxic, or less than 25 tons per year of a combination of listed toxic air pollutants. The Act requires that EPA determine within five years which area sources pose the greatest health risk. Once these particular sources have been identified, they must be regulated. An EPA study to identify and control those area sources posing the greatest health risk must be developed and published by November 1995. Controls for these smaller sources may be as stringent as the MACT controls for major sources, but more flexible measures called "Generally Available Control Technology (GACT)" will be used in some cases.

Finally, EPA is also required to establish a list of 100 or more hazardous substances that are particularly hazardous to human health and the environment when inadvertently released into the air by an unanticipated or uncontrolled event. Facilities that use or process these substances over established threshold quantities will be required to prepare risk management plans and comply with additional pollution prevention regulations.[126]

1.7.4 Ozone Depleters

A third type of air pollutant regulated by the Act includes the emissions of substances that deplete the upper (stratospheric) ozone layer. This depletion exposes life on earth to very harmful ultraviolet radiation. Facilities that repair and maintain air conditioning equipment are major source of these emissions.

---

[126] See Dennison, OSHA and EPA Process Safety Management Requirements: A Practical Guide for Compliance (New York: Van Nostrand Reinhold, 1994).

Chapter 2

State Environmental Law - Overview

2.1   STATE POLLUTION CONTROL LAWS - GENERALLY

In addition to the federal environmental laws discussed in Chapter 1, companies and individuals must also know and understand the requirements of various state environmental laws and regulations which control pollution and environmental quality within state boundaries. The state pollution control laws in large part mirror and elaborate on federal environmental programs. These state laws usually have additional and/or different requirements from their federal counterparts and it is especially important to know the environmental reporting and recordkeeping requirements for your state. Obviously, it is beyond the scope of this book to provide more than an overview of state environmental law. This chapter is designed to provide a general understanding of state environmental programs, with particular emphasis on state laws governing hazardous substance cleanup, hazardous waste management, underground storage tank regulation, and property transfer environmental disclosures. The reader is advised to contact state environmental authorities, trade organizations, and the state Bar Association for detailed guidance on state-specific environmental laws and regulations.[1]

---

1 For a good presentation of state environmental laws, see Selmi and Manaster, State Environmental Law (Clark Boardman Callaghan, 1989 and Supps.)

2.2 HAZARDOUS SUBSTANCE CLEANUP LAWS

Most states have enacted environmental cleanup laws that closely parallel the CERCLA liability scheme discussed in Chapter 1. Like CERCLA, these state laws provide a mechanism for the government and, in most cases, private parties to recover environmental cleanup costs from responsible parties. Certain important aspects of some state laws do, however, differ from CERCLA.[2] For instance, some state laws do not provide for an innocent landowner defense to environmental liability. Moreover, the environmental cleanup laws of some states impose additional restrictions not found in CERCLA.

2.2.1 Illustrative Summaries for Selected States

In order to avoid or minimize environmental liability and ensure regulatory compliance with state law requirements concerning hazardous substance cleanup, an individual or company must first consult the applicable state statute and regulations. Table 2A in section 2.2.2 references each state's hazardous substance cleanup law and provides a good starting point for determining what the state-specific requirements are. In addition to this useful reference table, a summary of the hazardous substance cleanup laws of several different states is provided here.

California

The California Hazardous Substance Account Act provides the state with authority to conduct response and remedial actions.[3] California has established two separate cleanup funds, i.e., the Hazardous Substance Account and the Hazardous Substance Cleanup Fund, to finance the containment and cleanup of releases of hazardous substances where potentially responsible parties fail to undertake remedial action. The state may initiate cost recovery actions against potentially responsible parties pursuant to the statute's strict liability standard.[4] If a liable party can show that it is only responsible for a portion of the costs, then the party is only responsible for that portion. California expressly adopts and incorporates the class of liable parties that are set forth in CERCLA.[5] The California law departs from CERCLA's joint and several liability scheme by using a proportional liability standard for damages. A court will apportion costs using equitable principles. For those potentially responsible parties who submit to binding arbitration, the Hazardous Substances Cleanup Arbitration Panel will apportion liability for remedial action costs.[6] The law also provides for recovery

---

2 See McElfish and Pennington, Reauthorizing Superfund: Lessons from the States (Washington, D.C.: Environmental Law Institute, Dec. 1994).
3 Cal. Health & Safety Code art. 5, 25350-25359.
4 Cal. Health & Safety Code art. 5, 25363.
5 Cal. Health & Safety Code art. 5, 25323.5.
6 Cal. Health & Safety Code art. 5, 25356.3.

of punitive damages, and gives private claimants an explicit right of recovery against other parties for cleanup costs and personal injuries.

Connecticut

The Connecticut Emergency Spill Response Fund provides state authority for state remedial action pursuant to CERCLA. Connecticut has no independent CERCLA-related law providing for state cleanup and cost recovery. The Emergency Spill Response Fund provides monies for various purposes, including containment, removal, or mitigation of toxic and hazardous substance spills and releases; removal of hazardous wastes that the state Commissioner of the Department of Environmental Protection deems a threat or potential threat to human health or the environment; and payment of the state's share of costs of remedial action under CERCLA.[7]

Illinois

Under the Illinois Environmental Protection Act, the state provides for a Hazardous Waste Fund to expend resources and to take necessary preventive or remedial action whenever there has been a release or substantial threat of a release of hazardous substances.[8] The Illinois Hazardous Substances Pollution Contingency Plan effectuates the state's authority and responsibility to take preventive or remedial action.[9] The state may recover its cleanup costs; however, the law does not provide for cost recovery by private party claimants.[10] Although not identical, the Illinois Environmental Protection Act defines potentially responsible parties in a similar manner to CERCLA. The Act does not address the standard of liability, however, strict liability is implied. The Act provides for assessment of treble damages against any person who is liable for a release or threatened release of a hazardous substance but fails to undertake removal or remedial action without sufficient cause.[11]

Indiana

The Indiana Hazardous Waste Act creates a Hazardous Substance Response Trust Fund to control, contain and isolate any hazardous substance released within the state.[12] Liability under the state law is taken from CERCLA in that any person that is liable under CERCLA Section 107(a) "is liable, in the same manner and to the same extent to the state."[13] Since the state law incorporates the liability provisions of CERCLA by reference, liable parties are the same as those set forth in

---

7 Conn. Gen. Stat. 22a-451(d).

8 Ill. Ann. Stat. ch. 111 1/2, 1022.2.

9 Ill. Admin. Code tit. 35, 750.101.

10 Ill. Ann. Stat. ch. 111 1/2, 1022.2(f).

11 Ill. Ann. Stat. ch. 111 1/2, 1022.2(k).

12 Ind. Code 13-7-8.7-2.

13 Ind. Code 13-7-8.7-8(c).

CERCLA. Accordingly, liability for hazardous substance cleanup and response costs is strict and joint and several.[14] The Indiana Hazardous Waste Law does not address private cost recovery actions.

Iowa

Under Iowa's Environmental Quality Act, the state has created a Hazardous Substance Remedial Fund for remedial action concerning the release of hazardous substances.[15] The state may recover cleanup and response costs, as well as natural resource damages, from liable parties. A liable party is defined as one "having control over a hazardous substance.[16] Iowa's law expressly provides for strict liability.[17] In addition, if a violation of the law is willful, the person is liable for treble damages.[18] The law does not expressly provide for private cost recovery actions.

Michigan

The Michigan Environmental Response Act provides the state with authority to respond to hazardous substance releases and to recoup cleanup costs from responsible parties.[19] Although the responsible parties governed by Michigan's law closely parallel CERCLA's liability provisions, the law uses broader definitions of the terms "owner" and "operator." Michigan's law defines "owner" as a person who owns a facility; operator is defined as a person who is in control of or responsible for the operation of a facility.[20] The scope of liability extends to all owners and operators at the time of disposal, at present, or at any intervening time.[21] Liability is strict and joint and several. However, damages may be apportioned among responsible parties if responsible parties can show that the harm is divisible and should be apportioned. The Michigan Environmental Response Act also contains detailed provisions for contribution among liable parties.[22] The law provides for private cost recovery actions.[23]

Minnesota

The Minnesota Environmental Response and Liability Act provides for an Environmental Fund for responding to releases of hazardous substances or hazardous wastes.[24] Civil actions may be pursued to recover monies

---

14 Ind. Code 13-7-8.7-8.
15 Iowa Code 455B.423.
16 Iowa Code 455B.381(8).
17 Iowa Code 455B.392, 455B.396
18 Iowa Code 455B.392.
19 Mich. Comp. Laws Ann. 299.609, 299.610.
20 Mich. Comp. Laws Ann. 299.603(t) and (u).
21 Mich. Comp. Laws Ann. 299.612.
22 Mich. Comp. Laws Ann. 299.612c.
23 Mich. Comp. Laws Ann. 299.615.
24 Minn. Stat. Ann. 115B.17, 115B.20.

expended from the Fund.[25] Liability for costs incurred for environmental cleanup and response in connection with a release of hazardous substances is strict and joint and several.[26] Liability may be apportioned among responsible parties if a responsible party can show that costs should be apportioned.[27]

Missouri

Missouri's Hazardous Waste Management Law establishes a Hazardous Waste Remedial Fund which may be used to respond to hazardous waste emergencies and abandoned hazardous waste sites.[28] The state may recover reasonable cleanup costs from responsible parties.[29] Any person having control over a hazardous substance is strictly liable under the Act.[30] Missouri's law does not provide for a private right of action to recover cleanup and response costs from other responsible parties.

New Jersey

The New Jersey Spill Compensation and Control Act established a New Jersey Spill Compensation Fund for use in cleanup up hazardous substance releases.[31] The law was enacted four years before CERCLA and is probably the most advanced state hazardous substance cleanup law. It goes a step further than CERCLA by regulating releases of petroleum.[32] Responsible parties are strict and joint and severally liable for cleanup and response costs.[33] Private parties may recover their response costs from other liable parties.[34]

Ohio

Although Ohio does not have a law that parallels CERCLA, the state has set up a general fund that may be used for hazardous waste cleanup.[35] Thus, Ohio state law lacks its own independent hazardous substance liability provisions similar to CERCLA. Still, if the state Director of Environmental Protection determines that a hazardous or solid waste facility poses a substantial threat to public health, the Director is required to initiate appropriate action to abate the threat of contamination, including remedial action financed from the cleanup fund. The state may recover its costs from liable parties.36

---

25 Minn. Stat. Ann. 115B.17, 115C.04.

26 Minn. Stat. Ann. 115C.04, 115B.05.

27 Minn. Stat. Ann. 115C.08.

28 Mo. Rev. Stat. 260.391, 260.480.

29 Mo. Rev. Stat. 260.500.

30 Mo. Rev. Stat. 260.530, 260.500(8).

31 N.J. Stat. Ann. 58:10-23.11.

32 N.J. Stat. Ann. 58:10-23.11b(k).

33 N.J. Stat. Ann. 58:10-23.11g(a).

34 N.J. Stat. Ann. 58:10-23.11k - 58:10-2311n.

35 Ohio Rev. Code Ann. 3734.28.

36 Ohio Rev. Code Ann. 3734.20.

Oregon

The Oregon Hazardous Waste Remedial Action Act provides state remedial action authority to respond to releases of hazardous substance and oil.[37] Oregon has two funds for use in financing cleanup of hazardous substance spills and releases when liable parties fail to take remedial action. The Oil and Hazardous Material Emergency Response and Remedial Action Fund is available for cleanup of unauthorized disposal or discharge of hazardous waste, hazardous substances, and oil.[38] The Hazardous Substance Remedial Action Fund may be used to fund state remedial action for hazardous substance releases, as well as to pay for the state-cost share of CERCLA-authorized cleanups.[39] Under the Oregon Hazardous Waste Remedial Action Act, strict liability is imposed on responsible parties.[40] Any person may seek contribution from any other liable or potentially liable party.[41]

Pennsylvania

Pennsylvania's Hazardous Sites Cleanup Act provides for a state Hazardous Sites Cleanup Fund to provide monies to undertake emergency response and remedial action regarding releases and threatened releases of hazardous substances into the environment.[42] Responsible parties are strictly and jointly and severally liable for all response costs, as well as damages for injury to or destruction of natural resources, and any interest accruing thereon.[43] Moreover, the statute provides for liability without proof of causation for all damages within 2,500 feet of the perimeter of the area where a release has occurred.[44] The Pennsylvania Hazardous Sites Cleanup Act provides for private cost recovery actions.[45]

---

[37] Or. Rev. Stat. 466.540, 466.565, 466.590.
[38] Or. Rev. Stat. 466.675.
[39] Or. Rev. Stat. 465.380.
[40] Or. Rev. Stat. 466.567, 466.640.
[41] Or. Rev. Stat. 465.325.
[42] 35 Pa. Cons. Stat. Ann. 6020.901, 6020.902.
[43] 35 Pa. Cons. Stat. Ann. 6020.702.
[44] 35 Pa. Cons. Stat. Ann. 6020.1109.
[45] 35 Pa. Cons. Stat. Ann. 6020.1115.

2.2.2 Quick Reference Table of State Laws

TABLE 2A - STATE HAZARDOUS SUBSTANCE CLEANUP LAWS	
State	**Hazardous Substance Cleanup Law**
AL	Ala. Code 22-30A-3
AK	Alaska Stat. 46.08.010
AZ	Ariz. Rev. Stat. Ann. 49-282, 49-904
AR	Ark. Stat. Ann. 8-7-410
CA	Cal. Health & Safety Code 25330, 25385.3
CO	Colo. Rev. Stat. 25-16-104, 29-22-106
CT	Conn. Gen. Stat. 22a-451(d)
DE	Del. Code Ann. tit. 7, 9105, 9113
DC	D.C. Code Ann. 6-701 - 6-714
FL	Fla. Stat. 403.725
GA	Ga. Code Ann. 12-8-68
HI	Haw. Rev. Stat. 128D-4
ID	Idaho Code 39-4417
IL	Ill. Ann. Stat. ch. 111 1/2, 1022.2
IN	Ind. Code 13-7-8.7
IA	Iowa Code 455B.423
KS	Kan. Stat. Ann. 65-3452
KY	Ky. Rev. Stat. Ann. 224.876, 224.877
LA	La. Rev. Stat. 30:2205
ME	Me. Rev. Stat. Ann. tit. 38, 1364(b)
MD	Md. Health-Envtl. Code Ann. 7-218, 7-220
MA	Mass. Gen. L. ch. 21E
MI	Mich. Comp. Laws Ann. 299.609 - .610
MN	Minn. Stat. Ann. 115B.17, 115B.20
MS	Miss. Code Ann. 17-17-29(4) 260.391, 260.480
MT	Mont. Code Ann. 75-10-704

TABLE 2A Continued	
NV	Nev. Rev. Stat. 459.530
NH	N.H. Rev. Stat. Ann. 147-B:3. 147-B:6
NJ	N.J. Stat. Ann. 58:10-23.11i
NM	N.M. Stat. Ann. 74-4-8, 74-9-34
NY	N.Y. Envtl Conserv. Law 27-0916, 27-1313
NC	N.C. Gen. Stat. Ann. 130A-310.11
ND	N.D. Cent. Code 37-17.1-07.1(3)
OH	Ohio Rev. Code Ann. 3734.28
OK	Okla. Stat. tit. 63, 1-2018
OR	Or. Rev. Stat. 465.380, 466.675
PA	35 Pa. Cons. Stat. Ann. 6020
RI	R.I. Gen. Laws 23-19.1-23
SC	S.C. Code Ann. 44-56-160
SD	S.D. Codified Laws Ann. 34A-12-3
TN	Tenn. Code Ann. 68-46-205
TX	Tex. Health & Safety Code Ann. 361.276(a)
UT	Utah Code Ann. 26-14d-301
VT	Vt. Stat. Ann. tit. 10 1283, 6615, 8008
VA	Va. Code Ann. 10.1-1400
WA	Wash. Rev. Code 70.105D.030
WV	W. Va. Code 20-5G-5
WI	Wis. Stat. Ann. 144.442(9)
WY	Wyo. Stat. 35-11-901

2.3   HAZARDOUS WASTE MANAGEMENT LAWS

Most states have enacted hazardous waste management laws that closely parallel RCRA. Like RCRA, these state laws provide a framework for controlling the generation, storage, and disposal of hazardous wastes. Certain important aspects of some state laws do, however, differ from RCRA. For instance, some states use their own hazardous waste manifest to track wastes and many states impose additional reporting and recordkeeping requirements on waste generators. Penalties for state violations may also be tougher under state law. Some states have specific standards governing the location of hazardous waste management facilities.

2.3.1 Illustrative Summaries for Selected States

In order to avoid or minimize environmental liability and ensure regulatory compliance with state law requirements concerning hazardous waste management, an individual or company must first consult the applicable state statute and regulations. Table 2B in section 2.3.2 references each state's hazardous waste management cleanup law and provides a good starting point for determining what the state-specific requirements are. In addition to this useful reference table, a summary of the hazardous waste management laws of several different states is provided here.

California

The California Hazardous Waste Control Act parallels RCRA; however, it contains a number of additional requirements. For example, the law has additional hazardous waste listing criteria and additional penalty provisions, and imposes additional reporting requirements on those who handle hazardous wastes. The state law prohibits underground injection disposal of certain wastes and imposes additional permit requirements for disposal of "extremely" hazardous wastes. California's law also requires a minimum 2,000-feet buffer zone surrounding regulated facilities.[46]

Connecticut

The Connecticut Hazardous Waste Law generally follows the RCRA hazardous waste management framework but imposes some different requirements.[47] For example, the law contains different criteria for identification of hazardous wastes, places additional reporting requirements on waste generators, and regulates oil recovery facilities. Connecticut's law has its own waste manifest, creates additional notice requirements, and has facility location standards. Underground injection of hazardous wastes is prohibited.

Illinois

The Illinois hazardous waste management law closely parallels RCRA but imposes some significant additional requirements. For example, the state has its own hazardous waste manifest form and tracking requirements, plus additional hazardous waste recordkeeping and reporting requirements. Illinois also has facility siting requirements, and its own permit rules and civil penalty scheme.[48]

---

46 Cal. Health & Safety Code art. 9, 25202.5.
47 Conn. Gen. Stat. 22a-449.
48 415 ILCS 5/20.

Indiana

The Indiana Hazardous Waste Act closely mirrors the federal scheme, although it does impose some additional reporting requirements, and civil and criminal penalties.[49] The Indiana law also requires that facility owners and operators provide an environmental site assessment before siting a facility.

Iowa

The Iowa Hazardous Waste Management Act closely parallels RCRA and uses the Uniform Federal Manifest for tracking hazardous wastes. Iowa has stricter facility location standards, its own permit modification standards, and additional penalty provisions.[50]

Michigan

For the most part, Michigan's hazardous waste law follows RCRA, although it does have some substantial differences. The state uses its own manifest, imposes significant additional reporting requirements, and applies substantial penalties against violators.[51]

Minnesota

Minnesota's hazardous waste law closely parallels RCRA and uses the Uniform Federal Manifest for tracking hazardous wastes. The state law lists some additional hazardous wastes, provides siting requirements, has groundwater monitoring rules, and imposes strict penalties of up to $100,000 per violation.

Missouri

.The Missouri Hazardous Waste Management Law is similar to RCRA in many respects but it contains some additional requirements. For example, Missouri lists waste oil as hazardous, and has its own waste identification requirements, its own manifest system, and additional reporting requirements. This state law also provides for additional recordkeeping and monitoring requirements and requires that disposal facilities pay permit fees. Violators of the Missouri law are subject to substantial civil and criminal penalties.[52]

New Jersey

The New Jersey hazardous waste law is similar to RCRA but imposes additional reporting and financial responsibility requirements on

---

49 Ind. Code 13-1-12-1.
50 Iowa Code Ann. 455B.301.
51 Mich. Comp. Laws Ann. 299.501.
52 Mo. Rev. Stat. 260.425.

facility owners and operators. The New Jersey law uses the Uniform Federal Manifest but requires that some additional information accompany the form.

Ohio

The Ohio Solid and Hazardous Waste Disposal Law mirrors RCRA and uses the Uniform Federal Manifest to track hazardous wastes. The Ohio law imposes additional facility siting requirements, and adds permit requirements, reporting requirements, and penalties.[53]

Oregon

The Oregon Solid and Hazardous Waste Control Law closely parallels RCRA and uses the federal manifest form to track hazardous wastes. The Oregon law imposes some additional reporting requirements and penalties.

Pennsylvania

The Pennsylvania Solid Waste Management Act is similar to RCRA but it has a few additional requirements. For example, the state uses its own hazardous waste manifest and imposes some additional labeling requirements on regulated wastes. The state law also adds some reporting requirements and penalties.[54]

2.3.2 Quick Reference Table of State Laws

TABLE 2B - STATE HAZARDOUS WASTE MANAGEMENT LAWS	
State	**Hazardous Waste Management Law**
AL	Ala. Code 22-30-1
AK	Alaska Stat. 46.09.020
AZ	Ariz. Rev. Stat. Ann. 49-901
AR	Ark. Stat. Ann. 82-4213
CA	Cal. Health & Safety Code art. 9, 25202.5
CO	Colo. Rev. Stat. 25-15-101
CT	Conn. Gen. Stat. 22a-449
DE	Del. Code Ann. 6301
DC	None
FL	Fla. Stat. Ann. ch.

---

53 Ohio Rev. Code Ann. 3734.01.
54 35 Pa. Cons. Stat. Ann. 6018.101.

TABLE 2B Continued	
GA	Ga. Code Ann. 12-8-60
HI	Haw. Rev. Stat. 342J-11
ID	Idaho Code 39-4401
IL	415 ILCS 5/20
IN	Ind. Code Ann. 13-1-12-1
IA	Iowa Code Ann. 455B-301
KS	Kan. Stat. Ann. 65-3441
KY	Ky. Rev. Stat. Ann. 224.994
LA	La. Rev. Stat. 30:2151
ME	Me. Rev. Stat. Ann. tit. 38, 1301
MD	Md. Health-Envtl. Code Ann. 8-1413, 8-1414
MA	Mass. Gen. L. ch. 21E,
MI	Mich. Comp. Laws Ann. 299.501
MN	Minn. Stat. Ann.
MS	Miss. Code Ann. 17-17-29
MO	Mo. Rev. Stat. 260.425
MT	Mont. Code Ann. 75-10-417
NE	Neb. Rev. Stat. 81-1501
NV	Nev. Rev. Stat. 459.400
NH	N.H. Rev. Stat. Ann. 1905.09
NJ	N.J. Stat. Ann.
NM	N.M. Stat. Ann. 74-4-1
NY	N.Y. Envtl Conserv. Law 27-0101
NC	N.C. Gen. Stat. Ann. 130A-290
ND	N.D. Cent. Code 23-20.3
OH	Ohio Rev. Code Ann. 3734.01
OK	Okla. Stat. tit. 63, 1-2001
OR	Or. Rev. Stat.
PA	35 Pa. Cons. Stat. Ann. 6018.101

TABLE 2B Continued	
RI	R.I. Gen. Laws 23-19.1
SC	S.C. Code Ann.
SD	S.D. Codified Laws Ann. 34A-11
TN	Tenn. Code Ann. 68-46-101
TX	Tex. Health & Safety Code Ann. 361.001
UT	Utah Code Ann.
VT	Vt. Stat. Ann. tit. 10, 6601
VA	Va. Code Ann. 10.1-1400
WA	Wash. Rev. Code Ann.
WV	W. Va. Code
WI	Wis. Stat. Ann. 144.43
WY	Wyo. Stat. 35-11-501

2.4   UNDERGROUND STORAGE TANK REGULATION

Most state underground storage tank programs parallel the framework provided in RCRA Title I with provisions for performance standards, financial responsibility, and corrective action. Most states, however, have more stringent penalty provisions and the list of regulated tanks often differs from the list contained in the RCRA regulations. Since a discussion of each state's UST law is not possible here, the reader should consult the applicable state law referenced in Table 2C on the next page. State-specific reporting requirements for UST spills and releases are, however, examined in detail in Chapter 4.

2.4.1 Quick Reference Table of State Laws

TABLE 2C - STATE UNDERGROUND STORAGE TANK LAWS	
State	**Underground Storage Tank Law**
AL	Ala. Code 22-30-1
AK	Alaska Stat. 46.03.826
AZ	Ariz. Rev. Stat. Ann. 49-1013
AR	Ark. Stat. Ann. 82-4215(e)
CA	Cal. Health & Safety Code 25297.1(h)(2)
CO	Colo. Rev. Stat. 8-20-503
CT	Conn. Gen. Stat. ch. 446K, 22a-416
DE	Mirror federal RCRA regs.
DC	None
FL	Fla. Stat. Ann. ch. 376.30 - 376.319
GA	Ga. Code Ann. 12-13-12, 12-13-19
HI	Haw. Rev. Stat. 342-61 - 342-71
ID	Idaho Code 39-4401
IL	Ill. Ann. Stat. 111 1/2, 1022.2(f)
IN	Ind. Code 13-7-20-21
IA	Iowa Code 455B.471
KS	Kan. Stat. Ann. 65-34105
KY	Ky. Rev. Stat. Ann. 224.60-135
LA	La. Rev. Stat. 30:2195
ME	Me. Rev. Stat. Ann. tit. 32, 10001
MD	Md. Health-Envtl. Code Ann. 4-409
MA	Mass. Gen. L. ch. 21J
MI	Mich. Comp. Laws Ann. 299.701
MN	Minn. Stat. Ann. 116.49
MS	Miss. Code Ann. 49-17-413
MO	Mo. Rev. Stat. 319.129

TABLE 2C Continued	
MT	Mont. Code Ann.
NE	None
NV	None
NH	N.H. Rev. Stat. Ann. 146-C:9
NJ	N.J. Stat. Ann. 58;:10A-22
NM	N.M. Stat. Ann. 74-6B-6
NY	N.Y. Envtl Conserv. Law 17-1001, 37-0105
NC	N.C. Gen. Stat. Ann. 143-215.94
ND	N.D. Cent. Code 23-20.3-04.1
OH	Ohio Rev. Code Ann. 3734.01
OK	Okla. Stat. tit. 17, 309
OR	Or. Rev. Stat. 466.770, 466.825
PA	35 Pa. Cons. Stat. Ann. 6021.702 -.706
RI	R.I. Gen. Laws 46-12-3(u)
SC	S.C. Code Ann. 44-2-70
SD	S.D. Codified Laws Ann. 34A-2-99
TN	Tenn. Code Ann. 68-215-115
TX	Tex. Water Code Ann. 26.3513
UT	Utah Code Ann. 19-6-419, 19-6-423
VT	Vt. Stat. Ann. tit. 10, 1922
VA	Va. Code Ann. 62.1-44.34:8
WA	Wash. Rev. Code 90.76
WV	W. Va. Code 20-5H-21
WI	Wis. Stat. Ann. 101.143
WY	Wyo. Stat. 35-11-1414

2.5   PROPERTY TRANSFER ENVIRONMENTAL DISCLOSURES

In addition to state laws that parallel or reinforce CERCLA, RCRA, and other federal environmental laws, many states have enacted statutes specifically designed to address environmental concerns involved in real estate transactions. Several states have developed specific regulatory initiatives to eliminate the sale or transfer of property with serious environmental contamination.

2.5.1 Illustrative Summaries for Selected States

This section summarizes the property transfer environmental disclosure requirements of various states. Table 2D lists the property transfer environmental disclosure requirements for every state.

California

In California, owners of nonresidential real property must provide written notice to a buyer of any release of a hazardous substance that the owner knows or has reason to believe may be on or below the property.[55] Willful failure to provide notice can result in civil penalties of up to $5,000. The act also imposes similar reporting requirements on lessees of both residential and nonresidential property. Failure to notify the property owner can result in a unilateral termination of nonresidential leases. However, the law is broad-ranging and undefined. It fails to identify clearly what and how much constitutes a release, when notice must be given, and what constitutes a sale. As a result, many lessees, buyers, and sellers of property in California have adopted a conservative approach and are requiring environmental assessments to ensure that proper disclosures are made.

Connecticut

The Connecticut Transfer Act[56] requires the transferor of a facility generating more than 100 kilograms per month of hazardous waste or that handles, uses, transports or disposes of hazardous waste generated by others to deliver to the transferee a notarized certification that there has been no discharge of hazardous waste on the site or that any such discharge has been cleaned up in accordance with state law. The certification must further state that any remaining hazardous waste is handled appropriately. A copy of the certification must be filed with the Connecticut Department of Environmental Protection within 15 days after the transfer. If the transferor cannot give the required certification, one of the parties to the transaction must certify to the state that it will remediate the contamination under the supervision of the state. The transactions that trigger a compliance obligation include:

---

55 Cal. Health & Safety Code 25359.7.
56 Conn. Gen. Stat. Ann. 22a-454(b).

* Change in ownership
* Sale of the stock or a merger or consolidation of a corporate owner
* Sale of controlling assets of the facility
* Conveyance of the real property
* Change in corporate identity or a financial reorganization

A corporate reorganization not substantially affecting ownership of the facility does not trigger a compliance obligation, nor does a cessation of operations. The statute provides a penalty of up to $100,000 for submitting false information. Failure to comply renders the transferor strictly liable for all cleanup and removal costs and all direct and indirect damages of the transferee.

Illinois

Under the Illinois Responsible Transfer Act,[57] a transferor is required to provide a transferee and any lender associated with the transaction with a disclosure statement, containing information specified in the statute, not more than 30 days after execution of a contract and not less than 30 days prior to the transfer. If the transferor fails to provide the disclosure statement, or if the disclosure statement reveals previously unknown environmental defects, the transferee or lender has the right to avoid any obligation to accept or finance the transfer, but only prior to the closing. The transferee or lender may not void the transfer after closing. However, an action for damages will lie, and an award of attorneys fees and costs may be made, in the discretion of the court. The transferor is also required to file the disclosure statement with the county and the Illinois Environmental Protection Agency.

The statute covers property with a facility that is required to comply with the federal Emergency Planning and Community Right to Know Act. The statute also reaches property with underground storage tanks that are required to be registered under the Resource Conservation and Recovery Act. The statute excludes from its coverage those portions of the property that are not the subject of the transfer.

A transfer includes:

* Lease of 40 or more years
* Transfer of a greater than 25% interest in a land trust
* Transfer of a power of direction of a land trust
* Deed or other instrument of conveyance

---

57 Ill. Ann. Stat. Ch. 30, para. 901.

Transfers that are not covered include:

* Corrections and modifications to an existing ownership structure
* Tax transfers
* Property used as security for a debt
* Ownership changes resulting from foreclosure of mortgages or liens
* Conveyances of mineral, oil and gas interests.

The transferor is subject to a $1,000 per day penalty for failure to provide the transferee with a disclosure document, a $10,000 per day penalty for submitting false information, and a $10,000 per day penalty for failure to record the disclosure statement.

Indiana

Indiana has enacted a disclosure law similar to the Illinois statute. The Indiana Environmental Hazardous Disclosure and Responsible Party Transfer Law,[58] enacted in 1989 and amended in 1990, requires a transferor to provide to transferees and lenders a disclosure statement with respect to environmental defects prior to the transfer of property.

The statute reaches many transfers, including:

* Conveyances of fee title
* Leases for more than 40 years, including all exerciseable option periods
* Assignments of more than 25% of the beneficial interest in a land trust and a collateral assignment of a beneficial interest in a land trust
* Installment contract
* Mortgage or deed of trust
* Lease of any duration, which includes an option to purchase.

Transfer does not include:

* Instrument to correct a previously recorded deed or trust document or which changes title without changing beneficial ownership
* Tax deed or a deed from a county pursuant to state law
* Deed of partition
* Conveyance resulting from foreclosure of a lien or a deed in lieu of foreclosure
* Easement
* Conveyance of a mineral, oil or gas interest
* Inheritance, devise or a conveyance by operation of law to a surviving joint tenant
* Foreclosure of a collateral assignment of a beneficial interest in a land trust
* Deed conveying fee title under an installment contract

---

58 Ind. Code 13-7-22.5-1.

"Transferor" includes a seller, grantor, mortgagor, or lessor of real property; an assignor of more than a twenty-five percent interest in a land trust; and the owners of a beneficial interest in a land trust. "Transferee" includes the buyer, mortgagee, grantee, or lessee of real property; an assignee of more than a twenty-five percent interest in a land trust; and the owners of a beneficial interest in a land trust. The terms transferor and transferee also include prospective transferors and transferees.

The transferor is required to deliver a disclosure statement to each of the other parties to a transfer of property, including the lender, 30 days before the transfer. The 30-day period may be waived in writing by the parties; however, the disclosure document must be delivered on or before the date on which the transfer is to become final. If the disclosure document reveals an "environmental defect" that was previously unknown to the other party, then the other party is relieved of any obligation to accept the transfer of the property or finance the transfer of the property. If the transferor fails to deliver the disclosure document, the party who did not receive the disclosure document may demand it and, if it is not received within 10 days or if, upon receipt, it discloses an environmental defect, may void an obligation to accept or finance the transfer of the property. A party to a transfer may not void an obligation to accept or finance the transfer of property after the transfer has taken place; however, a transferor is liable for all of a transferee's consequential damages.

The form of the disclosure document is prescribed by the statute.[59] The form starts out with a warning to the parties to the transfer that it is highly unlikely that reading the disclosure document would constitute sufficient due diligence to qualify for the "innocent purchaser" defense under CERCLA. The statute requires the disclosure document to be recorded in the county recorder's office. The transferor and transferee are jointly responsible for recording. If the recorded disclosure document reported the existence of an environmental defect, a person who has a financial interest in the property may record a document, certified by a registered professional engineer who does not have a financial interest, which reports that the environmental defect has been eliminated from the property.

A preprinted form is set out in the statute, which requires disclosure of the following information:

1. *Property Identification:* a legal description of the property;
2. *Liability Disclosure:* transferor and transferee of real property are advised that their ownership or other control of such property may render them liable to environmental cleanup costs whether or not

---

59 Ind. Code 13-7-22.5-15.

they caused or contributed to the presence of environmental problems in association with the property.

3. *Property Characteristics*:lot size; use (apartments, commercial, stores, industrial, farm, other);

4. Nature of the transfer;

5. Identification of the transferor;

6. Identification of the transferee;

7. *Environmental information*:

* Has the transferor ever conducted operations on the property that involved the generation, manufacture, processing, transportation, treatment, storage, or handling of a hazardous substance?

* Has the transferor ever conducted operations on the property that involved the processing, storage, or handling of petroleum other than that associated with the transferor vehicle use?

* Has the transferor ever conducted operations on the property that involved the generation, transportation, storage, or treatment of hazardous wastes?

8. Specific uses like landfill, surface impoundment, land application, waste pile, incinerator storage tanks, container storage area, injection wells, wastewater treatment plants, septic tanks, transfer stations, waste recycling operations, waste treatment detoxification, or disposal areas warrant further explanation on the form.

Iowa

The Iowa Environmental Quality Act[60] provides that the owner of an abandoned or uncontrolled disposal site listed on the state registry of such sites must apply for written approval from the state Department of Natural Resources, Environmental Protection Division, for the sale, conveyance, or transfer of title or a substantial change of use of the site. The Department then issues an approval or disapproval with respect to the sale, conveyance, transfer of title, or change of use.

Michigan

Whenever a person has actual or constructive knowledge or information that his property is a facility at which a hazardous substance has been released in a reportable quantity under section 299.610a(1)(c) of the Michigan Environmental Response Act, he may not transfer any interest in the property unless he notifies the purchaser or other transferee in writing that the property is a facility and discloses the general nature and extent of the release. The written notice provided by the transferor must be a separate written instrument from the instrument conveying the real property interest. If the instrument conveying the interest in real property is recorded, the written notice must be recorded with the register of deeds in the same county. Upon completion of all response

---

60 Iowa Code Ann. 455B.430.

activities for a facility, as approved by the state Department of Natural Resources, the owner of real property for which the above-described notice has been recorded, may record a certification with the register of deeds that all required response activity has been completed.[61]

Minnesota

Before transferring ownership of any property that the owner knows, or reasonably should know, contains hazardous waste, the owner must record an affidavit with the recorder of the county where the property is located, disclosing to any potential transferee that such property contains hazardous waste.[62]

Further, Minnesota adopted legislation in 1989 which is designed to require the seller of real property to notify the purchaser of any known wells on the property. Under the Minnesota Well Disclosure Law,[63] the seller of real property must provide the purchaser with a legal description of the property and a map showing the location of any known wells on the property. In addition, a well certificate must be filed with the deed. The County Recorder is not permitted to accept a deed without a well certificate.

Missouri

Under the Missouri Solid Waste Law,[64] the owner of an abandoned or uncontrolled disposal site that is listed on the state registry of such sites must notify the transferee in connection with any sale, conveyance, change of use, or transfer of title that the property is on the state registry. The owner must notify the transferee of all restrictions on the use of the site. The owner must also notify the transferee of potential liability for the site and notify the state regulatory agency within 30 days of the transaction.

New Jersey

New Jersey's Industrial Site Recovery Act (ISRA) is one of the most stringent state environmental laws to impact the sale or transfer of real estate. In 1993, ISRA amended and replaced New Jersey's ten-year-old Environmental Cleanup and Responsibility Act (ECRA).[65] ISRA is designed primarily to provide for expeditious cleanup of industrial properties at the time they are closed, sold, or otherwise change ownership. ISRA defines the types of properties subject to regulation

---

61 Mich. Comp. Laws Ann. 299.610c.

62 Minn. Stat. Ann. 115B.16.

63 Minn. Stat. 1031.235.

64 16 Mo. Stat. Ann. 260.465.

65 See 1993 NJ Sess. Law Serv. Ch. 139 (Senate 1070) (West), amending N.J. Stat. Ann. tit. 13, ch. 1K.

based on Standard Industrial Code (SIC) classification. The law provides exclusions for residential and many types of commercial properties. ISRA imposes cleanup responsibility upon the present owner or operator of an industrial site and mandates that (1) the site be free of contamination or (2) an approved remedial action plan be approved prior to the transaction. There are strict sanctions associated with the failure to comply with ISRA.

ISRA was enacted in 1993 to address criticisms that the environmental cleanup restrictions imposed by ECRA were too stringent. ALthough ECRA had become a national model in requiring environmental auditing and cleanup prior to the sale or closure of industrial properties, no other state enacted a similar law that was as tough as ECRA. ISRA attempts to streamline the remediation process, ease cleanup-related financial constraints on business, and make the administrative process associated with environmental audits more predictable. In enacting ISRA, the state legislature attempted to address the major criticisms of ECRA by:

* Allowing expedited environmental reviews at industrial sites that have undergone state-approved cleanups in the past
* Permitting businesses to defer remediation of an industrial property that is being transferred if the new owner will use the property for substantially the same purpose
* Allowing most soil cleanups to proceed without prior approval from state environmental regulators while requiring greater state oversight of ground water and surface water cleanups
* Permitting a property owner to transfer ownership of up to one-third of the value of an industrial site, and an even larger portion under certain conditions, without triggering a mandatory cleanup of the entire site
* Making allowances for the use of caps, fences, use restrictions, and similar engineering and institutional controls as alternatives to permanent remediation of a site

The most controversial aspect of the new law is the authority it gives the New Jersey Department of Environmental Protection and Energy to establish different cleanup standards based on exposure risk at a given site. A lower standard will be applied to nonresidential sites than to residential sites, on the assumption that people are less exposed to contamination on a nonresidential property.

Ohio

On June 29, 1994, Ohio Gov. George V. Voinovich signed the Ohio Real Estate Re-Use and Cleanup Law.[66] The new legislation takes a somewhat unique approach in that it sets up a voluntary program to allow cleanups to proceed without intense government oversight and, if done correctly, would release the owner from further liability. The Ohio EPA has

66 Ohio Rev. Code 3746.26.

identified at least 1,100 eligible sites. Sites that are covered under RCRA's underground storage tank regulations, are on the CERCLA National Priorities List, or are under enforcement action by the state are not eligible for the new program. The major provisions of this new legislation include:

* *Established cleanup standards.* Ohio EPA would develop protective standards designed to suit the intended use of the industrial property. Cleanups for residential use would have tougher standards than for industrial use.

* *Liability protection.* Banks and trustees would be released from environmental liability. Liability would be limited for local governments and contractors involved in voluntary cleanups. This protection is expected to free up necessary funding and encourage contractors and local governments to take part in cleanup projects.

* *Certification and auditing.* Ohio EPA would certify consultants and laboratories that want to do business with property owners who are performing voluntary cleanups. Anyone participating in a voluntary cleanup would have to use the services of Ohio EPA-certified professionals. The agency would also be given the right and responsibility to audit laboratories, records of certified professionals, and cleanup sites to make sure that they are conforming to specific standards.

* *Property Revitalization Board.* The law would establish a Property Revitalization Board made up of directors from various state agencies to serve as a clearinghouse for information on economic and financial incentives available to persons undertaking voluntary cleanup actions. The board would also review cleanup proposals from property owners who cannot remediate the property to cleanup standards due to extreme financial or technological hardship.

* *Cost Recovery.* The law would allow property owners to recover cleanup costs from other parties that contributed to the contamination at the site.

* *Consolidated permitting.* The law would streamline the permitting process by requiring only one permit that encompasses all environmental protection requirements, thus simplifying and speeding up cleanup efforts.

* *Tax Abatement.* To provide incentives for cleanup of contaminated property, the law would abate taxes for five years on the increased value of property from cleanup under the program.

Oregon

In Oregon, the owner of property must disclose to a potential purchaser information concerning any releases of hazardous materials that took place during the period that the seller owned the property.[67] If the owner or operator has actual knowledge of the release of

---

[67] Oregon Hazardous Waste Removal and Remedial Action Act, Or. Rev. Stat. 465.200-220 and 465.900.

hazardous materials on its property during its ownership or operation, it must disclose that fact to a potential purchaser. Failure to disclose renders the owner or operator strictly liable for remedial actions incurred by the state or any other person and for damages, including natural resource damages. This liability appears to be substantially the same as the owner-operator liability under CERCLA.

Pennsylvania

Pennsylvania has enacted two laws that require notice in connection with the conveyance of property that has been exposed to hazardous substances. The Pennsylvania Solid Waste Management Act[68] provides that the grantor in every deed for the conveyance of property on which hazardous waste is presently being disposed of (or has ever been disposed of) must include an acknowledgement concerning the hazardous wastes in the property description section of the deed. The acknowledgment must include, to the extent available, information about the service area, size and exact location of the disposed waste, and a description of the types of hazardous wastes found on the property. This acknowledgment in the property description is required as part of the deeds for all future conveyances or transfers of the property; however, the warranty in the deed does not apply to the information in the hazardous waste acknowledgment.

The term "hazardous waste" is defined in the definition section of the Solid Waste Management Act.[69] The definition is extremely broad and covers most wastes. However, "hazardous waste" specifically excludes coal refuse, including treatment sludges from coal mine drainage treatment plants, disposal of which is carried out pursuant to and in compliance with a valid permit.

The Pennsylvania Hazardous Sites Act[70] requires that a site at which hazardous substances remain after completion of a response action may not be put to a use that would disturb or be inconsistent with the response action. The Pennsylvania Department of Environmental Protection is obligated to require the Recorder of Deeds to record an order pursuant to the Act in a manner that will ensure its disclosure in the ordinary course of a title search. Such an order under the Act, when recorded, is binding on subsequent purchasers. The Act also requires the grantor in a deed of conveyance to record a notice similar to the notice required under the Pennsylvania Solid Waste Management Act.

---

68 35 Pa. Cons. Stat. Ann. 6018.405.

69 35 Pa. Cons. Stat. Ann. 6018.103.

70 35 Pa. Cons. Stat. Ann. 6020.512.

2.5.2 Quick Reference Table of State Laws

TABLE 2D - STATE PROPERTY TRANSFER ENVIRONMENTAL DISCLOSURES	
State	**Property Transfer Environmental Disclosure Law**
AL	Ala. Admin. Code r. 335-14-5-.07
AK	Alaska Stat. 46.03.822
AZ	Ariz. Admin. Code R18-8-265
AR	None
CA	Cal. Health & Safety Code 25230, 25359.7
CO	6 Colo. Code Regs 264.12(c), 264.116, 264.120
CT	Conn. Gen. Stat. 22a-134 - 22a-134d
DE	Del. Hazardous Waste Regs. 264.119
DC	None
FL	None
GA	None
HI	None
ID	None
IL	Ill. Ann. Stat. ch. 30
IN	Ind. Code 13-7-22.5
IA	Iowa Code 455B.426
KS	Kan. Admin. Regs. 28-29-20
KY	Ky. Rev. Stat. Ann. 224.876(16)
LA	La. Rev. Stat. 30:2157(A)
ME	*License Law and Rules Reference Book of the Real Estate Commission of the State of Maine,* ch. 19 (1988)
MD	Md. Regs. Code 26.13.05.07
MA	Mass. Gen. L. ch. 21E, 15
MI	Mich. Comp. Laws Ann. 299.610c
MN	Minn. Stat. Ann. 115B.16

TABLE 2D Continued	
MS	None
MO	Mo. Rev. Stat. 260.465, 260.470
MT	Mont. Code Ann. 75-10-715(6)
NE	None
NV	None
NH	None
NJ	N.J. Stat. Ann. 13:1K-8, 13:1K-9, 13:1K-10
NM	N.M. Stat. Ann. 74-9-24(F)
NY	N.Y. Envtl Conserv. Law 27-1305
NC	N.C. Gen. Stat. Ann. 130A-310.8
ND	N.D. Cent. Code 33-24-05-03, 33-24-05-68
OH	Ohio Rev. Code Ann. 3734.12, 3734.22
OK	Okla. Stat. tit. 63, 1-2004
OR	Or. Rev. Stat. 466.375
PA	35 Pa. Cons. Stat. Ann. 6018.405
RI	None
SC	None
SD	S.D. Admin. R. Ann. 74:27:15:07
TN	Tenn. Comp. R. & Regs. ch. 1200-1-11-.06
TX	Tex. Health & Safety Code Ann. 361.168
UT	Utah Admin. R. 450-7-14
VT	None
VA	None
WA	Wash. Admin. Code 173-303-610(10)
WV	W. Va. Code 20-5E-20
WI	None
WY	None

PART II ENVIRONMENTAL REPORTING AND RECORDKEEPING

Chapter 3

Environmental Reporting and Recordkeeping Requirements

3.1   INTRODUCTION

Companies that generate, store, handle, and dispose of hazardous substances must comply with various federal and state environmental reporting and recordkeeping requirements. Whenever a hazardous substance is released into the environment, the company must act quickly to contain it, thoroughly document the release, and report it to the appropriate federal, state, and local authorities. Generally, the Resource Conservation Recovery Act (RCRA);[1] Comprehensive Environmental

---

1 See 3.3.

Response, Compensation, and Liability Act (CERCLA);[2] the Emergency Planning and Community Right-to-Know Act (EPCRA);[3] the Clean Water Act (CWA);[4] and the Toxic Substances Control Act (TSCA)[5] contain the most important federal provisions governing reporting of spills and releases of hazardous substances, hazardous wastes, and oil.[6] State laws generally have similar and additional reporting and recordkeeping requirements. To obtain information on state requirements, check with your state hazardous waste/materials management authority.[7] State reporting procedures are summarized in Chapter 4.

3.2 SPILL NOTIFICATION AND RESPONSE - GENERALLY

Generally, each of the federal environmental laws has its own set of reporting requirements, and more than one law may be applicable to the reporting of a particular incident. Moreover, most of these laws require immediate notifications to regulatory authorities whenever a hazardous substance is released into the environment. It is essential that a company plan ahead to save valuable time in trying to determine which reporting requirements are applicable in a given situation. Table 3A summarizes the spill notification requirements of various federal environmental laws.

TABLE 3A - SPILL NOTIFICATION REQUIREMENTS

Statutory Provision	Implementing Regulation
Resource Conservation and Recovery Act (RCRA)	
42 U.S.C. 6921	
[] Spill, fire, explosion, or release of "hazardous waste" by a generator	40 C.F.R. 262.34
42 U.S.C. 6924	
[] Release, fire, or explosion at a treatment, storage, or disposal facility (TSDF)	40 C.F.R. 254.56
[] Statistically significant evidence of contamination during groundwater sampling	40 C.F.R. 264.98-.99
[] Tank system or secondary containment system leak or spill	40 C.F.R. 264.196

---

2 See 3.4.

3 See 3.5.

4 See 3.6.

5 See 3.7.

6 Spill notification requirements are summarized in Table 3A.

7 See Appendix B for a listing of state agency addresses and phone numbers concerning regulation of hazardous materials and wastes.

<u>42 U.S.C. 6925</u>

[] Release, fire, or explosion at an interim status TSDF	40 C.F.R.	265.56
[] Significant increase in hazardous waste constituents during groundwater sampling at an interim status TSDF	40 C.F.R.	265.93
[] Release to the environment from a tank system or secondary containment system at an interim status TSDF	40 C.F.R.	265.196

<u>42 U.S.C. 6991c</u>

[] Release from an underground storage tank (UST)	40 C.F.R.	280.61
[] Suspected release from a UST	40 C.F.R.	280.50
[] Spill or overfill of a UST	40 C.F.R.	280.53

Comprehensive Environmental Response, Compensation and Liability Act (CERCLA)

<u>42 U.S.C. 9603(a)</u>

[] Release of a "hazardous substance" in a "reportable quantity" (RQ)	40 C.F.R.	302.6

<u>42 U.S.C. 9603(f)</u>

[] "Continuous release" of RQ of a "hazardous substance"	40 C.F.R.	302.8

Emergency Planning and Community Right-to-Know Act (EPCRA)

<u>42 U.S.C. 11004(a), (b)</u>

[] Release of "an extremely hazardous substance" or a CERCLA "hazardous substance"	40 C.F.R.	355.40

Clean Water Act (CWA)

<u>33 U.S.C. 1321(b)</u>

[] Discharge of oil from a vessel, offshore or onshore facility	40 C.F.R.	110.10
[] Discharge of RQ of "hazardous substance" from a vessel, offshore or onshore facility	40 C.F.R.	117.21

<u>33 U.S.C. 1317</u>

[] Substantial change in discharge to a POTW	40 C.F.R.	403.12

Toxic Substances Control Act (TSCA)

<u>15 U.S.C. 2605(e)</u>
[] PCB spills   40 C.F.R. 761.125

Timely reporting of spills may prove especially difficult because many of the notification requirements are based on "immediate" or "twenty-four hour" notice. Compounding the problem, releases may occur during nonbusiness hours when information is hard to obtain. Further, facts may be distorted by employees who are fearful of blame or by decisionmakers who are under tremendous stress during a crisis situation. In addition, detailed reports are often required before all the relevant facts have been determined properly. Finally, it may be unclear to decisionmakers exactly which notification requirements must be fulfilled.

Ignorance is, however, no defense when it comes to compliance with the specific notification requirements of each federal environmental law. To assist in compliance, the following checklist outlines some common procedures that can be followed when reporting spills and releases of hazardous substances. These general guidelines must be used in conjunction with the specific reporting requirements of each environmental law.

3.2.1 General Checklist for Reporting Spills and Releases

DECIDE WHETHER TO REPORT
[] Evaluate information concerning the incident
  [] substance released
  [] quantity released
  [] location of release
  [] when release occurred
[] Decide whether there is a legal duty to notify
  [] evaluate any reporting requirements incorporated
     into permits, orders or decrees
  [] examine statutory or regulatory reporting
     obligations
     - "release"
     - "regulated substance"
     - "reportable quantity"
     - release to "environment"
     - *exemptions* from reporting requirements

DECIDE WHEN AND HOW TO REPORT
[] Immediate notification required?
[] Which government authorities must be notified?
  [] National Response Center
  [] EPA Regional Administrator
  [] Local Emergency Planning Commission
  [] State Emergency Response Commission

[] What proper mode of communication is permitted?
   [] telephone communication
   [] written notification

DECIDE WHAT TO REPORT
[] Chemical name and identity of any substance released
[] Indication whether substance is "extremely hazardous"
[] Estimate of quantity of substance released.
[] Time and duration of release
[] Medium or media into which release occurred
[] Known or anticipated acute or chronic health risks associated with release
[] Precautions that should be taken in responding to release
[] Names and telephone numbers of contact persons

DECIDE WHETHER FOLLOW-UP REPORTING IS REQUIRED

3.3 RCRA REPORTING AND RECORDKEEPING

RCRA plays a key role in the regulation of hazardous waste spills and releases. Discharges and spills of a hazardous waste that endanger health or the environment must be reported to appropriate federal and state authorities within twenty-four hours. Companies should also have emergency response plans in place in the event of a release. When responding to a spill or release, the discharger should carry out the following measures:

1. Eliminate the source of the release or stop the release of material into the environment.
2. Assess the character, amount, and overall extent of the release.
3. Contain the release to minimize its impact on the environment.
4. Recover the spilled material, including any contaminated soils.

RCRA's release reporting provisions primarly affect permitted treatment, storage, and disposal (TSD) facilities, although hazardous waste generators are also required to report a spill, fire, explosion, or release of "hazardous waste."[8] The RCRA reporting requirements that most impact waste generators relate to underground storage tank releases.[9]

---

8 RCRA 3001, 42 USC 6921; 40 CFR 262.34.
9 See 3.3.1.

The owner or operator of a RCRA permitted TSD facility must:

1. Orally report within twenty-four hours of becoming aware of the circumstances, any noncompliance that may endanger health or the environment; and
2. Provide a written report within five days.[10]

More specifically, the owner or operator of the permitted facility must report any information concerning any hazardous waste release that may endanger public drinking water supplies to the EPA or delegated state agency. The permittee must also report any fire or explosion from a TSD facility that could threaten the environment or human health outside the facility.[11]

Parallel reporting requirements for RCRA permitted and interim status TSD facilities are contained in 40 C.F.R. Section 264.56(d)-(e) and 40 C.F.R. Section 265.56(d)-(e). Each of these sections requires that the facility owner or operator immediately notify the on-scene coordinator for the geographical area in which the facility is located or the National Response Center if an imminent or actual emergency situation occurs.

In addition to reporting spills and releases, the EPA has issued regulations specifying the types of records that must be kept by generators of hazardous wastes.[12] The following list shows the necessary retention periods and filing frequency for different RCRA records.

TABLE 3B - RCRA RECORDS

Type of Record	Retention Period
Copy of hazardous waste manifests	3 years
Test results	3 years
Waste analysis	3 years
Biennial reports	3 years
Manifest exception reports	Active life of facility
Training records	Until closure or 3 years after employee left

---

10 EPA may waive this five day reporting requirement and instead require submission of a written report within 15 days. 40 CFR 270.30(l)(6).

11 40 CFR 270.30(l)(6).

12 RCRA 3002(a)(1), 42 USC 6922(a)(1); 40 CFR 262.40-262.44.

Contingency plan   Active life
                                           of facility

EPA Identification Number   Active life
                                           of facility

3.3.1 Underground Storage Tank Releases

Owners and operators of underground storage tanks must report various conditions to the authorized environmental agency within twenty-four hours, including:

* Any spill or overfill of petroleum resulting in the release into the environment of an amount in excess of 25 gallons (or other reportable quantities of other regulated substances).[13]
* Discovery of a leak or release, including free product or vapor in the soil, sewers, or waters.
* Suspected releases must be investigated within 7 days. An investigation consists of tank tightness tests, a site assessment (soil sampling and analyses), or both, depending on the reason for suspecting a release.[14]
* Instances of malfunction or erratic behavior of UST equipment, unless the owner can confirm that the equipment is defective, but not leaking, and repairs the equipment immediately.
* Monitoring results that indicate a possible release, unless the monitoring device is found to be defective and immediately repaired, replaced, or recalibrated.[15]

Suspected releases other than the last two conditions will require a tank tightness test or site investigation.[16] Inventory discrepancies need not be reported until after a second month's observation confirms the existence of a possible leak.[17] A penalty of up to $25,000 per day may be imposed for failure to report UST releases.[18]

3.3.2 UST Corrective Action

Upon confirmation of a release, the owner or operator must:
1. Report the release to the authorized agency;
2. Take immediate action to prevent further release (for example, closing a leaking valve or removing product from the UST);
3. Remove all free product from the environment; and

---

13 40 CFR 280.53. The UST owner or operator must report any spill or overfill of a hazardous substance that results in a release to the environment in a quantity equaling or exceeding its reportable quantity as listed in 40 CFR Part 302.

14 40 CFR 280.52.

15 40 CFR 280.50.

16 40 CFR 280.50-280.53.

17 40 CFR 280.50(c)(2).

18 42 USC 6928, 6991e.

4. Mitigate fire, explosion, and vapor hazards.[19]

Within twenty days after confirmation, the owner/operator must report to the authorized agency, summarizing the initial response and providing any additional information.

After consultation with the agency, the owner/operator must perform a site assessment, develop a response plan, and complete remedial action. The remedial action may include aeration or replacement of contaminated soil and prevention of groundwater contamination.[20] The nature of remedial action depends largely on site-specific factors, such as the type and quantity of substance released, the proximity to surface and groundwater, and the type of soil. A release causing groundwater contamination becomes very expensive and may necessitate long-term groundwater treatment or provision of alternate sources of drinking water for affected residents. A owner/operator's liability for contamination of a municipal water supply could be financially devastating.

Most state UST programs do not require removal from the site of soil contaminated with petroleum in every case, but the owner/operator must manage or treat it to prevent or minimize environmental risk. The normal procedure for cleaning up small petroleum spills is to:

1. Dig up the contaminated soil;
2. Place the soil on an impermeable sheet to prevent contaminants from returning to the soil; and
3. Periodically turn the soil over to allow hydrocarbon contaminants to escape into the air.

During aeration, the contaminated soil should also be covered with a top plastic sheet if necessary to prevent rain water from leaching contaminants from the contaminated soil. After aeration has reduced the hydrocarbon levels below applicable levels, the soil may be returned to the excavation. As an alternative, contaminated soil can often be removed and transported to a municipal landfill for aeration and disposal, provided that the landfill is permitted to receive the type and levels of contamination in the soil requiring disposal.

More extensive or severe contamination will probably require more expensive remediation, including disposal of contaminated material in a hazardous waste or special waste landfill, treatment of groundwater, discharge of treated groundwater into the public sewer system, soil venting, incineration of contaminated materials, or biodegradation. Excavation and aeration or disposal of a small leak might cost up to $25,000 to $30,000. The cost of correcting more extensive leaks

---

19 40 CFR 280.61-280.63.
20 40 CFR 280.65-280.66.

requiring biodegradation or treatment of groundwater could easily exceed
$100,000.

3.3.3 Underground Storage Tank Records

There are a series of recordkeeping requirements for underground
storage tanks. Records may be kept on-site for immediate inspection by
the authorized agency, or may be kept in a readily available alternative
site, such as corporate or division headquarters, for inspection upon
request by the agency.[21] In some states, decals or annual permits must
be affixed to the fill-pipe or displayed at the UST site.

The list below shows the various reports that must be filed with the
authorized environmental agency, as well as other records that must be
retained. In states with an approved UST program, the authorized
environmental agency is the state department responsible for
administering the program, while in other states, the authorized agency
is EPA.

TABLE 3C - UST RECORDS[22]

Reports to File With the Environmental Agency

* Notification of all new UST systems, including certification of
proper installation
* Reports of releases and suspected releases
* Corrective action planned or taken
* Notification of permanent closure or change-in-service

Records to Maintain on File

* A corrosion expert's analysis if corrosion protection is not
used
* Documentation that corrosion protection equipment has been
tested and is properly operating
* Documentation relating to repairs
* Compliance with release detection requirements
* Site investigations conducted at permanent closure

In addition to the required records, the owner should keep complete
and well-organized files of all documents relating to its USTs,
including any technical drawings, and photographs of all tank
installations. Files also should contain records of warranties and
performance claims, any schematics and blueprints, and all diagrams,
maps, or surveys of tanks, piping, and monitoring wells. These records
should be kept on hand at least until the tank is permanently closed. By
keeping detailed and organized files, the cost of subsequent repairs or

---

21 40 CFR 280.34(c).

22 40 CFR 280.34(b)(1)-(5).

retro-fits will be reduced, and faster response to emergencies and releases will be made possible.

3.4 CERCLA REPORTING AND RECORDKEEPING

Pursuant to the Comprehensive Environmental Response, Compensation and Liability Act (CERCLA), a company is obligated to report any hazardous waste release in any amount equal to or greater than the reportable quantities outlined in the regulations.[23] The company must immediately notify the National Response Center at (800) 424-8802. Appropriate state authorities must also be notified. Failure to notify the National Response Center immediately upon discovery of the reportable release exposes the company to fines of up to $25,000 per day of violation and/or imprisonment of up to three years. Delays of even one or two hours have resulted in penalties.[24] Repeat offenders face fines of up to $75,000 per day of violation and/or imprisonment of up to five years.[25]

In order to respond promptly in the face of the "immediate" notification of a spill or release of a hazardous substance pursuant to CERCLA Section 103(a), the company must know what hazardous substances and wastes it has at the site.[26] The CERCLA hazardous substances are listed at 40 C.F.R. Section 302.4 which provides the reportable quantity for each of the 725 listed hazardous substances.[27] If a substance is unlisted, the reportable quantity is generally 100 pounds or more.[28]

When determining the RQ for mixtures of hazardous and nonhazardous substances, only the hazardous component is considered. For example, for a 50% acetic acid solution, 10,000 pounds of the solution must be released to be reportable, since the RQ for acetic acid is 5,000 pounds.[29] If the mixture contains any regulated substance for which the amount released is unknown, notification is required when the total amount of the mixture released equals or exceeds the RQ for the hazardous constituent with the lowest RQ. One court has determined that

---

23 CERCLA 103(a), 42 USC 9603(a). The reportable quantities are listed in 40 CFR 302.4.

24 See In re Great Lakes Div. of Nat'l Steel Corp., Docket No. EPCRA-007-1991 (July 31, 1993) (EPA assessed $60,000 in penalties against company under CERCLA and EPCRA for alleged failure to report a release within two hours).

25 CERCLA 109, 42 USC 9609.

26 Note that because certain petroleum products are expressly excluded from the definition of "hazardous substance" in CERCLA 101(14) (the so-called "pertroleum exclusion"), releases of petroleum are not subject to CERCLA reporting requirements.

27 It is also important to note that the list in Table 302.4 is not a complete listing because it does not include certain ignitable, corrosive, or reactive substances that are RCRA characteristic hazardous wastes. 40 CFR 302.4(b).

28 40 CFR 302.5(b).

29 40 CFR 302.6(b)(1).

the RQ for soil containing an unknown amount of toluene was 1,000 pounds, which is the RQ for toluene when not contained in soil.[30]

3.4.1 Exemption for Continuous Releases

Releases of regulated substances in amounts less than the RQ within a twenty-four-hour period are NOT reportable under CERCLA. This temporal component is one basis under which gradual releases of hazardous substances have been excluded from CERCLA Section 103(a) notification.[31] By way of illustration, a slow leak of acetic acid from a storage tank at a rate of five pounds per hour would not be reportable under CERCLA. Still, this release might be reportable under under state hazardous substance regulations.

3.4.2 Other Exemptions

In addition to exempting "continuous" releases, CERCLA exempts three other types of releases from its reporting requirements:

1. Federally permitted releases, as defined in CERCLA Section 101(10);
2. Application of pesticide projects that are registered under the Federal Insecticide, Fungicide, and Rodenticide Act (FIFRA); and
3. Certain releases of hazardous waste that must be reported under RCRA and which are reported to the National Response Center.

3.5 EPCRA REPORTING AND RECORDKEEPING

Under the Emergency Planning and Community Right-to-Know Act (EPCRA), also known as Title III of the Superfund Amendments and Reauthorization Act (SARA),[32] facilities are required to immediately report accidental spills or releases of hazardous substances to a designated "Community Emergency Coordinator for the local emergency planning committees and to the state emergency planning commission for any areas likely to be affected by the release."[33] There are two classifications of hazardous substances that must be reported under EPCRA when released in threshold quantities. The first is when the substance is on EPA's published list of "extremely hazardous substances."[34] The second is when notice of a spill or release is required pursuant to CERCLA Section 103(a).[35]

Besides communication of spills and releases, regulated companies must also submit two different annual reports to the appropriate government authorities:

---

30 United States v. MacDonald & Watson Waste Oil Co., 933 F.2d 35 (1st Cir. 1991).

31 CERCLA 103(f)(2).

32 42 USC 11001-11050.

33 EPCRA 304(b), 42 USC 11004(b).

34 40 CFR Part 355, Appendix A.

35 42 USC 9603(a).

1. A hazardous chemical inventory required pursuant to EPCRA Section 312; and

2. A Toxic Chemical Release Inventory Form R required under EPCRA Section 313.

The hazardous chemical inventory requires each company to submit a Material Safety Data Sheet (MSDS) for each chemical found on its premises in the threshold quantity specified on EPA's published list of "extremely hazardous substances."[36] These MSDSs are identical to those required under the Occupational Safety and Health Act.[37] The chemical inventory must be filed on March 1 of each year on either a Tier One or Tier Two form.[38]

3.5.1 Emergency Reporting of Spills and Releases

Under EPCRA Section 304, a the company is required to report immediately accidental spills or releases of hazardous substances to a designated "Community Emergency Coordinator" for the LEPC and to the SERC for any areas likely to be affected by the release."[39] There are two categories of hazardous substances that must be reported:

1. Releases of EHSs at or above TPQs listed in Appendix A to 40 C.F.R. Section 355; and

2. Releases subject to reporting under CERCLA Section 103(a), 42 U.S.C. Section 9603(a), which requires a company to report any hazardous substance release in any amount equal to or greater than the reportable quantities (RQs) outlined in 40 C.F.R. Section 302.4.

If no reportable quantity is specified for a given substance, the RQ is one pound.

It is important to remember that the CERCLA definition of "hazardous substance" is quite broad and incorporates by reference RCRA hazardous wastes. Thus, although certain hazardous substances may not be found on the CERCLA list of hazardous substances in 40 C.F.R. Section 302.4, they may still be subject to emergency notification requirements. Thus, releases of RCRA "characteristic" hazardous wastes are subject to reporting under EPCRA. The RQ for such hazardous wastes is 100 pounds. However, if the waste exhibits the RCRA characteristic of toxicity, the RQ is the RQ for the toxic constituent in the waste. If the waste exhibits other hazardous characteristics (ignitability, corrosivity, or reactivity) in addition to toxicity, the RQ for the entire quantity of waste is the lowest RQ for any substance exhibiting any one of the

---

36 See 40 CFR 302, 304 for the EPA list.

37 29 USC 651-678.

38 See 3.5.5.

39 EPCRA 304(b), 42 USC 11004(b).

hazardous characteristics. Thus, for reporting purposes, the entire quantity of waste is considered to exhibit the characteristic of the constituent with the most reportable quantity.[40]

3.5.2 Reporting Exemptions

Several categories of releases are exempt from EPCRA emergency reporting requirements.[41] The company's emergency response coordinator must understand the reporting requirements for every environmental law, and know whether any exemptions apply. This is no easy task, and requires advance training and planning, but is essential to avoiding costly noncompliance penalties, as well as ensuring cost-effective environmental management. Knowledge of exemptions can, of course, prevent unnecessary alarm and administrative expense. The following releases are exempt from EPCRA Section 304 emergency reporting requirements:

* Federally permitted releases, as defined in CERCLA Section 101(10), including discharges authorized by RCRA permit; air emissions subject to a federal or state permit or control requirement; effluent discharges subject to a Clean Water permit; discharges to public sewer system if in compliance with pretreatment standards.

* "Continuous" releases which require annual reporting under CERCLA Section 103(f) and EPCRA Section 304, unless there is a "statistically significant increase" in the release, a new release, or a change in the "normal range" of the release.[42]

* Releases that are exempt from CERCLA reporting under CERCLA Section 101(22) (releases to persons solely within the workplace) or which result in exposures to persons solely within the facility site boundaries.[43] However, note that some releases - although occurring on-site - may cause off-site exposures, such as when contaminants seep into groundwater and migrate to off-site location.

* Application of a registered pesticide product that is exempt from CERCLA Section 103(e) reporting.

* Any radionuclide release that occurs naturally in the soil, naturally from the disturbance of land or from the dumping of coal or coal ash piles.

3.5.3 Emergency Notification Procedures

When a release requires notification under EPCRA Section 304, the company must provide an initial notification and a follow-up report. The initial notification may be made by calling 911 if it occurs incident to

---

40 40 CFR 302.5(b).

41 40 CFR 355.40(a)(2).

42 See 40 CFR 302.8(g).

43 EPCRA 304(a)(4).

transportation. Otherwise, the initial notification must be given immediately to the SERC and the LEPC, and must include the following information:

* Chemical identification;
* Whether the chemical is listed as an EHS;
* Estimated quantity of the release;
* Time and duration of the release;
* Media contaminated by the release;
* Known or anticipated acute health effects;
* Precautionary measures to be taken in response to the release;
* Name and telephone number of a contact person at the site;

Following verbal notification, a written report must be submitted to the SERC and LEPC as soon as practicable, and include the following information:

* Actions to be taken to respond to and contain the release;
* Any known or anticipated acute or chronic health effects associated with the release; and
* If appropriate, advice regarding medical attention necessary for any exposed individuals.[44]

EPA has stated that even if the facility owner/operator was unaware of the release he will not be relieved of the duty to report it if he "should have known" of the release.[45]

Obviously, there is going to be overlap in reporting requirements under EPCRA, CERCLA, and other environmental laws. Reporting a release under one law, however, does not relieve the regulated individual or company of reporting under another law. Thus, it is very important for the company or individual to know the reporting requirements of every environmental law. For example, release of a EHS above its TPQ would require immediate notification under EPCRA Section 304. If the EHS is also listed as one of the hazardous substances in 40 C.F.R. Section 302.4, the National Response Center must also be notified pursuant to CERCLA Section 103(a). Conversely, some of the EHSs listed in the EPCRA regulations are not also listed as hazardous substances under CERCLA. If there is a reportable release of an EHS under EPCRA that is not listed as a hazardous substance under CERCLA, reporting is technically only required to be made to the SERC and the LEPC.[46] EPA is, however, planning to designate all EHSs as CERCLA hazardous substances, which would result in multiple notification under EPCRA, CERCLA, and possibly under other environmental laws.

---

44 EPCRA 304(b),(c), 42 USC 11004(b),(c).

45 52 Fed. Reg. 13393 (Apr. 22, 1987).

46 See 52 Fed. Reg. 13386 (Apr. 22, 1987).

Noncompliance with the EPCRA Section 304 emergency release reporting provisions exposes violators to civil penalties of up to $25,000 a day for each separate violation, and $25,000 per day for each day of noncompliance. In cases of a second violation, civil penalties of $75,000 can be imposed for each day of noncompliance.[47]

3.5.4 Chemical Inventories

In addition to emergency notification of reportable spills and releases, EPCRA Sections 311, 312, and 313 require that certain companies submit chemical inventories to appropriate agencies. Under EPCRA Section 311, each company subject to OSHA's Hazard Communication Standard[48] must submit a Material Safety Data Sheet (MSDS) for each chemical found on its premises in an amount equal to or exceeding its TPQ. These MSDSs are identical to those required under the Occupational Safety and Health Act.[49] In lieu of submitting an MSDS for each chemical, the company may submit a list of chemicals for which MSDSs are required. If a list is provided, the chemicals must be grouped according to different "hazard categories":[50]

1. Acute health hazards;
2. Chronic health hazards, including carcinogens;
3. Fire hazards;
4. Sudden pressure release hazards; and
5. Reactivity.

The MSDSs or chemical list must be submitted to the SERC, the LEPC, and the local fire department.

OSHA exempts many chemicals from the Hazard Communication Standard and, therefore, from the chemical inventory requirements of EPCRA Sections 311 and 312.[51] EPCRA also exempts certain substances from the toxic chemical inventory requirements of Section 311 and 312.[52] It is important to note that those chemicals exempted from the inventory requirements of Sections 311 and 312 are not necessarily exempted from the emergency notification requirements of Section 304.

The minimum threshold quantity triggering reporting under EPCRA Sections 311 and 312 is either (1) 500 pounds or the TPQ for EHSs, whichever is less; or (2) 10,000 pounds for any other hazardous chemical.[53] Thus, to determine the TPQ, it is necessary to calculate the total amount, in pounds, of each hazardous chemical present on-site at

---

47 40 CFR 355.50(a)(b).

48 See 29 CFR 1910.1200.

49 29 USC 651-678.

50 40 CFR 370.2.

51 See 29 CFR 1910.1200(b)(6) for this list of exemptions.

52 See EPCRA 311(e), 312(c).

53 40 CFR 370.20(b)(1).

any given time and compare the actual amount of each substance with the aforementioned amounts.

Failure to submit chemical lists or MSDSs under EPCRA Section 311 may result in penalties of up to $10,000 per violation.[54]

3.5.5 Tier One and Tier Two Reporting Forms

Companies subject to the chemical inventory requirements of EPCRA Section 311 must also submit annual hazardous chemical inventory forms pursuant to EPCRA Section 312. This hazardous chemical inventory must be filed on March 1 of each year on either a Tier One or Tier Two form. All reporting on the Tier One form is done in the aggregate and describes total amounts of hazardous substances grouped by their hazardous properties.

The Tier Two form is more specific in that it requires quantity, location, and hazard information for each specific hazardous material on site. A company must submit a Tier Two form if it is requested by the state or local environmental authorities or by the local fire department. If desired, the Tier Two may be submitted in place of the Tier One.

Some companies may find it easier to complete this form instead of the Tier One since it basically demands the same information that must be gathered to complete the Tier One form. On the Tier Two form, instead of grouping chemicals by hazardous characteristic, the company provides information by individual chemical. The Tier One and Tier Two forms and instructions are found at 40 C.F.R. Sections 370.40(b) and 370.41(b), or may be ordered by calling the EPCRA hotline at (800) 535-0202.

Failure to submit Tier One forms can subject violators to penalties of $25,000 per day.[55]

3.5.6 Toxic Chemical Release Reporting (Form R)

Finally, in addition to the requirements of EPCRA Sections 311 and 312, certain companies must submit another annual report of releases of listed "toxic chemicals" pursuant to EPCRA Section 313, known as the Toxic Chemical Release Inventory (Form R). Regulated companies must be file this report on July 1 with the EPA and the relevant state authority. On Form R, the company reports any releases made during the preceding twelve months. Form R must be filed if the business has ten or more full-time employees, has a Standard Industrial Classification Code 20-39, and the business manufactures, stores, imports, or otherwise uses designated toxic chemicals at or above threshold levels. Generally, a company must file an annual Form R if it manufactures, imports, or

54 40 CFR 370.5(a).
55 40 CFR 370.5(b).

processes at least 25,000 pounds of a listed "toxic chemical," or if it uses at least 10,000 pounds of a listed "toxic chemical" during the previous calendar year. "Toxic chemicals" subject to the Form R reporting requirements and their respective threshold quantities are listed at 40 C.F.R. Section 372.65. Certain exemptions from the Form R reporting of toxic chemical releases may apply.[56] Violations of EPCRA Section 313 reporting are punishable by fines of up to $25,000 per day.[57]

EPA is also considering expanding the list of chemicals subject to Form R reporting. Other substances being considered for addition to the list include Clean Water Act priority pollutants; EHSs listed for purposes of EPCRA Section 312; RCRA "listed" hazardous wastes from 40 C.F.R. Sections 261.33(e), (f); and air toxins under Sections 112(b) and 602(a), (b) of the Clean Air Act Amendments of 1990.

With the passage of the Pollution Prevention Act of 1990 (PPA),[58] new requirements were added to the Form R. Section 6607 of the PPA expands and makes mandatory source reduction and recycling information on the EPCRA list of toxic chemicals. These requirements have been added by EPA through modifications to sections 6, 7, and 8 of the Form R. Section 6 has been modified so that off-site location and transfer amounts are reported together, including amounts sent off-site for recycling. Section 7 has been modified to include detailed information about on-site recycling activities, as well as changes to the information provided for treatment activities. Section 8 contains the majority of the new source reduction and recycling reporting requirements. For more information about the Form R, call the EPCRA hotline at (800) 535-0202.

3.6   CLEAN WATER ACT REPORTING AND RECORDKEEPING

Under Section 311(b)(5) of the Clean Water Act (CWA), any person that is "in charge of" a vessel or an onshore or offshore facility must immediately report a discharge of oil or hazardous substance into or upon navigable waters of the United States from such vessel or facility, when in quantities that may be harmful, as soon as such person has knowledge of the discharge.[59] Penalties of up to $25,000 per day may be imposed for failure to report releases as required under the Clean Water Act.[60]

EPA regulations at 40 C.F.R. Parts 116 and 117 implement the CWA Section 311 requirements concerning hazardous substances. The EPA list

---

56 See 40 CFR 372.38.

57 40 CFR 372.18.

58 42 USC 13101-13109.

59 CWA 311(b)(5), 33 USC 1321(b)(5).

60 33 USC 1321(b).

of CWA hazardous substances is found at 40 C.F.R. Section 116.4.[61] EPA has established specific reportable quantities (RQs) in 40 C.F.R. Section 117.3 for each of the hazardous substances so designated under the CWA.[62] When there is a release of a hazardous substance into navigable waters, a single report by the person in charge to the National Response Center will satisfy the spill notification obligations of both the CWA and CERCLA.[63]

3.6.1 Oil Discharges

An important aspect of the Clean Water Act's reporting requirements that differs from CERCLA is that oil discharges are reportable under the CWA.[64] Under the CWA, "oil" is defined as "oil of any kind or in any form, including, but not limited to, petroleum, fuel oil, sludge, oil refuse, and oil mixed with wastes other than dredged spoil."[65] The EPA's regulations governing the discharge of oil require immediate notification to the National Response Center as soon as the person in charge has knowledge of a release of oil in a quantity considered harmful.[66] A quantity is considered "harmful" when it violates applicable water quality standards for navigable waters[67] or when the oil causes a film or sheen on the surface water.[68]

3.6.2 Reporting Exemptions

The CWA contains a few general exemptions from its spill notification requirements. National Pollutant Discharge Elimination System (NPDES) permitted discharges are exempt,[69] as are discharges permitted under the International Convention for Prevention of Pollution of the Sea by Oil treaty.[70] Discharges of oil from properly functioning vessel engines are also exempted from reporting.[71]

---

61 Note, however, that this list is only a subset of the CERCLA list of hazardous substances found at Table 302.4 in 40 CFR Part 302.

62 These RQs are generally consistent with the RQs provided in 40 CFR 302.4.

63 See 51 Fed. Reg. 34340 (Sept. 29, 1986).

64 Such discharges are exempt from CERCLA reporting under the "petroleum exclusion" to CERCLA's definition of hazardous substance in CERCLA 101(14).

65 CWA 311 (a)(1).

66 40 CFR 110.10.

67 The definition of "navigable waters" has been interpreted to encompass virtually every body of water in the United States. See United States v. Zanger, 767 F. Supp. 1030 (N.D. Cal. 1993).

68 40 CFR 110.3. This latter standard is commonly known as the "sheen" regulation. See Chevron, U.S.A., Inc. v. Yost, 919 F.2d 27 (5th Cir. 1990).

69 CWA 311(a)(2).

70 CWA 311(b)(3)(A).

71 40 CFR 110.7.

3.7   TOXIC SUBSTANCES CONTROL ACT (TSCA)

Under Section 8(e) of the Toxic Substances Control Act (TSCA), any person who manufactures, processes, or distributes chemical substances or mixtures, and who obtains information which reasonably supports a conclusion that the substance or mixure presents a substantial risk of injury to health or the environment, must immediately inform the EPA of such risk.[72] To comply with TSCA Section 8(e) reporting requirements, a person with knowledge of the reportable event must telephone the EPA as soon as that person knows about the incident. In addition, a written follow-up report must be filed with the EPA within 15 days of the incident.[73]

EPA guidance states that this requirement applies to any environmental contamination by a "chemical substance" or "mixture" known to cause certain adverse effects, i.e, cancer, birth defects, death, or bioaccumulation, and which:

> 1. Seriously threatens humans with cancer, birth defects, mutation, death, or serious or prolonged incapacitation, or
> 2. Seriously threatens nonhuman organisms with large-scale or ecologically significant population destruction.[74]

The reporting requirements apply to a broad class of chemical substances and mixtures, which are defined in TSCA Section 3. According to the EPA guidance, reporting is only triggered when a person obtains information that "reasonably" supports a conclusion that such substances or mixtures pose a substantial risk of injury to health or the environment. In any case, reporting is not required if the person has actual knowledge that the EPA has already been notified or the incident has already been reported as a spill under CWA Section 311(b)(5).

Finally, EPA regulations implementing TSCA Section 6(e), which governs polychlorinated biphenyls (PCBs), require reporting of PCB spills.[75]

3.8   PENALTIES FOR NONCOMPLIANCE

The prospect of being hit with heavy fines and penalties should provide plenty of incentive for regulatory compliance. Failure to report spills and releases of hazardous substances into the air, ground, and water account for a significant portion of the penalties assessed by the

---

72 TSCA 8(e), 15 USC 2607(e).

73 See 52 Fed. Reg. 20083 (May 29, 1987).

74 See 58 Fed. Reg. 37735 (July 13, 1993); 43 Fed. Reg. 11110 (Mar. 16, 1978).

75 40 CFR 761.125.

EPA. In recent years, EPA and the Department of Justice (DOJ) have stepped up environmental enforcement activities.[76]

3.8.1 EPA/DOJ Enforcement Authority

EPA and DOJ are the two federal agencies primarily responsible for the enforcement of federal environmental laws.[77] Investigations are carried out by EPA's National Enforcement Investigations Center and DOJ's Federal Bureau of Investigation. If an investigation reveals a potential violation or if the government receives information by some other means (e.g., voluntary or mandatory reporting, employee allegations) concerning a violation of environmental laws, EPA and the DOJ have three general mechanisms for enforcing environmental laws: administrative actions, civil judicial actions, and criminal actions.

EPA may pursue administrative actions under various environmental statutes. Most major environmental statutes include provisions authorizing the EPA administrator both to order parties out of compliance with a statute to come into compliance with the statute,[78] and to impose monetary penalties for violations.[79]

Through DOJ, EPA may also bring civil judicial actions against alleged violators of every major environmental statute.[80] Generally, EPA conducts investigations of potential violators, determines that a violation is or may be occurring, and then refers the information to DOJ with a recommendation that DOJ bring a civil action against the alleged violator. DOJ has authority to make an independent decision concerning whether it should bring a civil action against the alleged violator. In most cases, DOJ attorneys appear in court on behalf of EPA. In a civil action, the government may seek both an injunction requiring the alleged violator to come into compliance with a law and a fine.

The last general type of formal action that EPA and DOJ may use to enforce an environmental law is a criminal judicial action. Like a civil judicial action, EPA refers potential criminal actions to DOJ, recommending that DOJ bring a criminal action. Criminal prosecutions of environmental laws have been on the rise in recent years.[81]

---

76 See "Enforcement Actions at EPA Continue to Climb in Civil, Criminal Cases, Penalty Assessments," 22 Env't Rep. (BNA) 1832 (Nov. 29, 1991).

77 State environmental agencies also possess enforcement powers for state environmental laws, and when EPA has delegated authority to a state to implement a federal program at the state level, the designated state agency will possess enforcement powers in conjunction with EPA oversight.

78 See RCRA 3008(a), 42 USC 6928(a).

79 See RCRA 3008(c), 42 USC 6928(c); CERCLA 109(a), 42 USC 9609(a).

80 See RCRA 3008(a)(1), 42 USC 6928(a); CERCLA 109(c), 42 USC 9609(c); EPCRA 325, 42 USC 11045.

81 According to DOJ statistics, over 34 years of prison confinement were imposed on individual defendants, and 103 corporate officers were indicted in fiscal year 1992. Department of Justice Memorandum, October 27, 1992.

3.8.2 EPA/DOJ Enforcement Activities

Civil and criminal prosecutions of violations are on the rise. In fiscal year 1994, the EPA took a record number of enforcement actions against violators of various federal environmental laws, including penalty assessments, use of supplemental environmental projects (SEPs), administrative consent decrees, and filing of civil and criminal cases. In FY94, EPA officials brought a total of 2,247 enforcement actions, including 220 criminal cases, and 1,597 administrative actions. These figures represent an increase of 137 actions over FY93.

In FY94, EPA referred 220 criminal cases for prosecution to the Department of Justice (DOJ), an increase of 36% over the previous fiscal year record of 140 DOJ-referred cases. EPA and DOJ brought criminal charges against 250 individual and corporate defendants, a 40% increase over last year. Almost $37 million in criminal fines were assessed (up from $30 million in FY93) and convicted defendants received a total of 99 years in jail sentences, a 25% increase over last 1993 figures.

3.8.3 Violators Learn Hard Lessons

Some companies have learned the hard way that the government is serious about enforcing its reporting requirements. In several cases, substantial fines and penalties were imposed on polluters for failure to report a spill or release of hazardous substances. For instance, in 1992, Exxon was fined a record $125 million for the *Valdez* oil spill, and Dexter Corporation was fined $4 million for violations of RCRA and the Clean Water Act.[82] In another case, a maintenance foreman at an Army installation instructed several workers to dispose of waste paint in a pit filled with water. Workers told him that the cans were leaking, but he continued to order disposal into the pit. He was later convicted and sentenced to one year of probation.[83]

In 1993, the Second Circuit Court of Appeals affirmed the decision of a federal district court convicting and sentencing a business owner to three years in prison for knowingly disposing of hazardous waste without a permit under RCRA and three years in prison for failing to report the release of hazardous substances as required under CERCLA. In addition to the two concurrent prison terms, the businessman was ordered to pay $607,868 in restitution.[84]

The defendant in the case, Harris Goldman, was co-owner of GCL Tie & Treating, Inc. (GCL), a railroad tie treating business located in Sidney, New York. (The named defendant Ken Laughlin was Vice President

82 No. H-89-393 (AHN) (D. Conn. Sept. 3, 1992).
83 United States v. Carr, 880 F.2d 1550 (2d Cir. 1989). See also U.S. v. Baytank (Houston), Inc., 934 F.2d 599 (5th Cir. 1991), wherein operators of a chemical transfer and storage facility were found guilty of failing to timely report the release of some 4,000 pounds of acrylonitrile.
84 United States v. Laughlin, 1993 WL 495765 (2nd Cir. Dec. 3, 1993).

of Operations and Plant Manager at GCL.) The tie treatment process consisted of first placing untreated green ties into a large cylinder and then adding creosote. The process generated creosote sludge, a hazardous substance.

When GCL began experiencing problems with its treatment process, excess creosote often became contaminated or spilled from the system. GCL supervisors regularly directed employees to dispose of the contaminated creosote by soaking it up with sawdust and dumping it in remote areas of the GCL property. On October 30, 1986, a large, accidental creosote spill occurred at GCL. Contrary to Goldman's instructions, this spill was reported to the New York Department of Environmental Conservation (DEC), which then began making periodic, pre-announced visits to the site.

In 1987, GCL began experiencing financial difficulties, which were magnified when GCL's boiler ceased to function properly. Without a properly functioning boiler, GCL could not recover creosote sludge left over from its treatment process. GCL quickly began to run out of storage space for the creosote sludge generated from the treatment process. GCL employees were directed to put the excess creosote sludge into an empty railroad tanker car that had recently brought a new shipment of creosote. Goldman, however, became concerned over GCL's daily accrual of rental charges for keeping the tanker car beyond its normal return date. After two weeks had passed with no resolution of the boiler problems, Goldman met with GCL's Vice President of Operations and Plant Manager, Ken Laughlin, and a company consultant to discuss possible methods of disposing of the creosote sludge. Goldman rejected two proposed legal means of disposing of the creosote sludge and began to demand repeatedly that Laughlin release the contents of the tanker car onto the ground. After Laughlin refused, Goldman himself released the entire contents of the tanker car directly onto the ground.

After GCL went bankrupt and the site was abandoned in 1988, DEC began to piece together details of the site condition. After a criminal investigation, Goldman was indicted, and later convicted, for illegal disposal activities at the site.

Goldman appealed his conviction on the grounds that the jury was improperly instructed on the knowledge component of the RCRA violation. Goldman contended that the government was required to prove beyond a reasonable doubt that he was aware of the applicable RCRA regulations applicable to creosote sludge and knew that GCL had not obtained a waste disposal permit. Initially, the court stated that it was only necessary that "a defendant have a general awareness that he is performing acts proscribed by the statute." On the issue of whether the defendant had to have knowledge of the lack of a permit, the court followed those jurisdictions (the majority rule) that have concluded that knowledge of the lack of a permit is not an requisite element of the offense under RCRA Section 3008(d)(2)(A), which provides criminal penalties for "[a]ny person who ... knowingly treats, stores, or disposes of any hazardous waste ... without a permit...."[85]

Goldman also contested the jury instructions on the CERCLA violation concerning his failure to report the release of creosote sludge. The court was equally unsympathetic to this argument, holding that CERCLA Section 103(a) does not require knowledge of regulatory requirements. The statute only demands that the defendant be aware of his actions.

Several other cases illustrate the foolish risk that violators of RCRA and Superfund have taken. For example, in *United States v. Greer*,[86] the owner and operator of a waste recycling and transportation business in Orlando, Florida, was found guilty under CERCLA for failing to report a 1,000-gallon release of a hazardous waste mixture containing approximately 80% 1,1,1-trichloroethane. Failure to report releases immediately to the National Response Center carries a possible fine of up to $25,000 per day of violation and/or imprisonment of up to three years. Repeat offenders face fines of up to $75,000 per day of violation and/or imprisonment of up to five years.

---

85 42 USC 6928(d)(2)(A).

86 850 F.2d 1447 (11th Cir. 1988).

Chapter 4

Environmental Reporting and Recordkeeping Procedures

4.1   INTRODUCTION

In response to the reporting and recordkeeping requirements of the various federal and state environmental laws and regulations, businesses and industrial facilities should have response plans in place in the event of a spill or release of pollutants in reportable quantities. The company's response plan should designate an emergency coordinator who knows the reporting requirements of each federal and state environmental law and who has been trained to follow specific emergency response procedures. The company's emergency response plan should be a written document with step-by-step procedures to follow.[1] It must be developed in accordance with regulatory compliance mandates and follow the reporting and recordkeeping procedures demanded by each environmental law. If regulatory requirements change, the plan should be updated and appropriate personnel should be kept apprised of all changes.

This chapter provides a good starting point for companies to develop their own environmental reporting and recordkeeping procedures. This chapter seeks, in practical fashion, to go through the essential reporting and recordkeeping procedures of the primary federal

---

1 See Section 5.2.

environmental laws. In addition to the federal procedures outlined, companies must be aware of all additional and/or different state reporting and recordkeeping requirements and procedures. It is not possible to give guidance here on the procedures for every state. However, a representative sampling of several state reporting and recordkeeping procedures is provided in Section 4.3.[2] Those in immediate need of state procedures not currently covered in this volume should contact their state environmental agencies for assistance.[3]

4.2 FEDERAL REPORTING AND RECORDKEEPING PROCEDURES

In the subsections that follow, general procedures are given to aid companies in complying with the federal reporting and recordkeeping requirements discussed in Chapter 3. It is important to bear in mind that a spill or release of pollutants in reportable quantities may trigger reporting under more than one federal environmental law, as well as additional state reporting requirements. Therefore, upon knowledge of a spill or release, the company must determine the substance discharged and the quantity released in order to determine which federal and state reporting requirements have been triggered, if any. Immediate action must be taken to contain the release and, depending on the quantity, immediate notification to federal and state authorities may be required. The procedures outlined here provide a systematic process for quickly determining:

1. When reporting is required and to whom;
2. Whether oral or written notification is necessary;
3. Information needed to complete a written report; and
4. Any follow-up procedures that may be required.

4.2.1 CERCLA - Reportable Hazardous Substance Releases

1. Reporting Trigger:

Any release of a hazardous substance into the environment equal to or exceeding the reportable quantity as measured over any 24-hour period. CERCLA 103(a); 42 U.S.C. 9603(a).

2. Reportable Substance:

"Hazardous substances" as defined in CERCLA 101(14), 42 U.S.C. 9601(14); 40 C.F.R Part 302.

3. Reportable Quantity (RQ):

Reportable quantities of hazardous substances can be found in Table 302.4 to 40 C.F.R. 302.

---

2 In future updates to this book, discussion of additional state procedures will be included.

3 State environmental agency contacts are listed in Appendix B.

4. Person Responsible for Reporting:

"Person in Charge."

5. Person/Agency to Notify:

National Response Center (24-hour) at (800) 424-8802 or (202) 267-2675.

6. Timing/Form of Notification:

Verbal notification immediately upon knowledge of a release of a RQ of a hazardous substance. A written report is not required.

The information reported should include:

- Owner of facility
- Location, quantity, and type of release
- Response action taken
- Nature of damage and injuries
- Whether state or local agency has been notified

7. Important Regulatory Definitions:

Environment: Environment is defined as all surface and ground water, land surface, or subsurface strata and ambient air within the United States or under the jurisdiction of the United States.

Hazardous Substance: The term "hazardous substance" is defined broadly in CERCLA by reference to substances defined as hazardous in a number of other environmental statutes, including the Resource Conservation and Recovery Act, 42 U.S.C. 6921, the Clean Water Act, 33 U.S.C. 1317(a), and the Clean Air Act, 42 U.S.C. 7412.

Release: The term "release" means any spilling, leaking, pumping, pouring, emitting, emptying, discharging, injecting, escaping, leaching, dumping, or disposing into the environment. Release also refers to a substantial threat of release.

Excluded from the definition of "release" is petroleum, including crude oil or any fraction thereof. The petroleum exclusion applies both to refined and unrefined gasoline, even though certain of its indigenous components and certain additives during the refining process have themselves been designated as hazardous substances.[4]

---

4 However, if petroleum is contaminated or adulterated through use, it is not protected by the petroleum exclusion.

8. Reporting Exclusions:

a. Continuous releases of hazardous substances that do not reach the reportable quantity limit during any 24-hour period *for which notice has already been given.*[5]

A release is considered continuous if it occurs without interruption or abatement or is routine, anticipated, and intermittent and incidental to normal operations or treatment processes.

b. Federally permitted releases, as defined in CERCLA Section 101(10);[6]

c. Application of pesticide projects that are registered under the Federal Insecticide, Fungicide, and Rodenticide Act (FIFRA); and

d. Certain releases of hazardous waste that must be reported under RCRA and which are reported to the National Response Center.

e. Releases to persons solely within the workplace or which result in exposures to persons solely within the facility site boundaries.[7]

f. Any radionuclide release that occurs naturally in the soil, naturally from the disturbance of land, or from the dumping of coal or coal ash piles.

9. Special Issues:

a. *Releases of mixtures or solutions of hazardous substances* (including hazardous waste streams) are subject to the following reporting requirements:

- If the quantity of all of the hazardous constituent(s) of the mixture of solution is known, notification is required where a RQ of any hazardous constituent is released; or
- If the quantity of one or more of the hazardous constituent(s) of the mixture or solution released equals or exceeds the RQ for the hazardous constituent with the lowest RQ.

---

5 CERCLA 103(f)(2). Note, however, that notification must initially be given to the National Response Center concerning continuous releases of a hazardous substances. Written notice must also be provided to the appropriate EPA Regional Office within 30 days of the telephone notification to the NRC.

6 "Federally permitted release" is defined by CERCLA to cover discharges which are in compliance with permits issued under certain federal environmental statutes. 42 USC 9601(10).

7 CERCLA 101(22), 42 USC 9601(22). However, note that some releases (although occurring on-site) may cause off-site exposures, such as when contaminants seep into ground water and migrate to off-site locations.

b. *Releases of mixtures or solutions containing radionuclides* must be reported under the following circumstances:

- If the identity and quantity (in curies) of each radionuclide in a released mixture of solution is known, the ratio between the quantity released (in curies) and the RQ for the radionuclide must be determined for each radionuclide. The only such releases subject to these reporting requirements are those in which the sum of the ratios for the radionuclides in the mixture or solution released is equal to or greater than one.

- If the identity of each radionuclide in a released mixture or solution is known but the quantity released (in curies) of one or more of the radionuclides is unknown, the only such releases that must be reported are those in which the total quantity (in curies) of the mixture or solution released is equal to or greater than the lowest RQ of any individual radionuclide in the mixture or solution.

- If the identity of one or more radionuclides in a released mixture or solution is unknown (or if the identity of a radionuclide released by itself is unknown), the only such releases subject to reporting requirements are those in which the total quantity (in curies) released is equal to or greater than either one curie or the lowest RQ of any known individual radionuclide in the mixture or solution, whichever is lower.

4.2.2 EPCRA - Reportable Hazardous Substance Releases

1. Reporting Triggers:

a. A release of an "extremely hazardous substance" into the environment equal to or exceeding its reportable quantity.[8]

·b. Notice of a spill or release is required pursuant to CERCLA Section 103(a).[9] Any release that must be reported to the National Response Center under CERCLA Section 103(a) must also be reported to *state and local planning committees*.

2. Reportable Substance:

a. *"Extremely hazardous substances"* as defined under EPCRA 304, which are listed in Appendices A and B of 40 C.F.R. Part 355.

b. *"Hazardous substances"* as defined in CERCLA 101(14), 42 U.S.C. 9601(14).

---

8 42 USC 11004; 40 CFR 355.40.

9 42 USC 9603(a).

3. Reportable Quantity (RQ):

a. The Reportable Quantity for each "extremely hazardous substance" is found in Appendices A and B of 40 C.F.R. Part 355.

b. The Reportable Quantity for each CERCLA "hazardous substances" is found in Table 302.4 to 40 C.F.R. 302

4. Person Responsible for Reporting:

The owner or operator of the facility.

5. Person/Agency to Notify:

Immediately upon occurrence of a reportable release notify the "Community Emergency Coordinator" for the Local Emergency Planning Committee (LEPC) and the State Emergency Response Commission (SERC) for any areas likely to be affected by the release.[10] Notification may be by telephone, radio, in person, or other similar means.[11] A written follow-up report must be submitted to the LEPC and SERC "as soon as practicable" after reporting of the release.[12]

6. Timing/Form of Notification:

a. Immediate oral notification must be given upon occurrence of a reportable release and must include the following information:

- The chemical name or identity of any substance involved in the release.
- An indication of whether the substance is an extremely hazardous substance.
- An estimate of the quantity of any such substance that was released into the environment.
- The time and duration of the release.
- The medium or media into which the release occurred.
- Any known or anticipated acute or chronic health risks associated with the emergency and, where appropriate, advice regarding medical attention necessary for exposed individuals.
- Precautions to take because of the release, including evacuation (unless such information is readily available to the community emergency coordination pursuant to the emergency plan).
- The names and telephone number of the person or persons to be contacted for further information.

b. The written follow-up notice required by 40 C.F.R 355.40(b)(3) must be submitted to the SERC. The written follow-up notice must

---

10 EPCRA 304(b), 42 USC 11004(b).

11 EPCRA 304(a), 42 USC 11004(a); 40 CFR 355.40(b)(1).

12 EPCRA 304(c), 42 USC 11004(c); 40 CFR 355.40(b)(3).

update the information previously reported and must include the following additional information:

- Actions taken to respond to and contain the release.
- Any known or anticipated acute or chronic health risks associated with the release.
- Where appropriate, advice regarding medical attention necessary for exposed individuals.

7. Important Regulatory Definitions:

Release: Release means any spilling, leaking, pumping, pouring, emitting, emptying, discharging, injecting, escaping, leaching, dumping, or disposing into the environment (including the abandonment or discarding of barrels, containers, and other closed receptacles) of any hazardous chemical, extremely hazardous substance, or CERCLA hazardous substance.

Environment: Environment includes water, air, and land and the interrelationship that exists among water, air, and land and all living things.

Facility: Facility means all buildings, equipment, structures, and other stationary items that are located on a single site or on contiguous or adjacent sites and that are owned or operated by the same person (or by any person who controls, is controlled by, or is under common control with, such person). For purposes of emergency release notification, the term includes motor vehicles, rolling stock, and aircraft.

8. Reporting Exclusions:

Certain releases are exempt from EPCRA reporting requirements.[13] The releases that are exempt from reporting requirements are those generally described in Section 4.2.1 above with respect to CERCLA Section 103(a).[14]

---

13 EPCRA 304(a), 42 USC 11004(a).
14 See CERCLA 101(22), 42 USC 9601(22).

4.2.3 RCRA - Reportable Hazardous Waste Discharges

1. Reporting Triggers:

a. *Treatment, storage, or disposal (TSD) facilities* must report any hazardous waste release from the facility which may endanger health or the environment.

b. *Small quantity generators* subject to RCRA regulation must immediately report any fire, explosion, or other release of hazardous waste which could threaten human health outside the facility. Notification is also required when the generator has knowledge that a spill has reached surface water.

2. Reportable Substance:

Hazardous waste, as defined under RCRA. 40 C.F.R. Part 261, "Identification and Listing of Hazardous Waste," contains the elements of the hazardous waste definition.

3. Reportable Quantity (RQ):

Amount of hazardous waste that could endanger health and the environment and/or create an emergency situtation.

4. Person Responsible for Reporting:

The owner or operator of the facility.

5. Person/Agency to Notify:

TSD Facilities:

a. Report hazardous waste releases from a TSD facility to the U.S. EPA or delegated state agency.

b. Report any fire or explosion from a TSD facility to local authorities if evacuation of local areas is advisable.

c. Report to the National Response Center (24-hour) at (800) 424-8802 or (202) 267-2675 if an imminent or actual emergency situation occurs.

Small Quantity Generators:

Report any fire, explosion, or other release of hazardous waste that poses a threat to human health outside the facility to the National

Response Center (24-hour) at (800) 424-8802 or (202) 267-2675 and the implementing state agency.[15]

6. **Timing/Form of Notification:**

TSD Facilities:

a. Orally report any information concerning the release from a TSD within 24 hours of knowledge of the release[16] or immediately if an emergency situation exists.[17]

b. Written follow-up notice to the U.S. EPA or delegated state agency is required within 5 days from the time that the facility owner or operator becomes aware of a reportable circumstance. The owner or operator must note in the operating record the time, date, and details of any incident that requires implementing a RCRA contingency plan.

The report must include:

- Name, address, and telephone number of the owner or operator.
- Name, address, telephone number, and U.S. EPA Identification Number of the facility.
- Date, time, and type of incident (e.g., fire, explosion).
- Name and quantity of hazardous wastes releases.
- The extent of injuries, if any.
- An assessment of actual or potential hazards to human health or the environment, where this is applicable.
- Estimated quantity and disposition of recovered wastes that resulted from the incident.

Small Quantity Generators:

Orally report any information concerning the release within 24 hours of knowledge of the release or immediately if an emergency situation exists. No written follow-up notice is required.

The information reported should include:[18]

- Name, address, and U.S. EPA Identification Number of the generator.
- Date, time, and type of incident.
- Quantity and type of hazardous waste.
- Extent of any injuries.
- Estimated quantity and disposition of recovered materials.

---

15 40 CFR 262.34(d)(5)(iv).

16 40 CFR 270.30.

17 40 CFR 264.56, 265.56.

18 40 CFR 262.34(d)(5)(iv).

7. Important Regulatory Definitions:

Release is not defined in the hazardous waste regulations. EPA generally uses the CERCLA definition of release.[19]

Facility means all contiguous land, structures, other appurtenances, and improvements on the land, used for treating, storing, or disposing of hazardous waste. A facility may consist of several treatment, storage, or disposal operational units (e.g., one or more landfills, surface impoundments, or combinations of them).

4.2.4 RCRA - Reportable UST Releases

1. Reporting Triggers:

A release or suspected release of a reportable quantity of petroleum or a CERCLA hazardous substance into the environment from an underground storage tank.

2. Reportable Substance:

Petroleum or CERCLA hazardous substances (40 C.F.R Part 302), excluding those that are hazardous wastes as defined under RCRA Subtitle C.

3. Reportable Quantity (RQ):

a. A spill or overfill of petroleum that results in a release to the environment exceeding 25 gallons or other reasonable amount specified by the implementing agency, or that causes a sheen on nearby surface water.[20]

b. A spill or overfill of a CERCLA hazardous substance that results in a release to the environment that equals or exceeds the RQ listed in 40 C.F.R Part 302.

4. Person Responsible for Reporting:

The owner or operator of the UST system.

5. Person/Agency to Notify:

Report to the U.S. EPA or state implementing agency.

---

19 See Section 4.2.1.

20 If the spill is less than 25 gallons, the release must be reported if it cannot be cleaned up within 24 hours.

6. Timing/Form of Notification:

a. *Petroleum releases:* Report to within 24 hours or other reasonable time specified by the implementing agency. Within twenty days after confirmation of the release, the owner/operator must report to the authorized agency, summarizing the initial response and providing any additional information.

Releases can be confirmed from:

- The presence of regulated substances in soils, ground water, basements, sewers and utility lines, nearby surface waters, or wells.
- Unusual operating conditions at the tank system, including sudden loss of a substance from a tank or the unexplained presence of water in a tank.
- Monitoring system results.

b. *Hazardous substance releases:* Report any release of a RQ to the environment from a hazardous waste tank system or secondary containment system within 24 hours of detection.[21]

Within 30 days of detection of a reportable release to the environment, a written report containing the following information must be submitted:

- Likely route of migration of the release.
- Characteristics of the surrounding soil (soil composition, geology, hydrogeology, climate).
- Results of any monitoring or sampling conducted in connection with the release (if available). If sampling or monitoring data relating to the release are not available within 30 days, these data must be submitted as soon as they become available.
- Proximity to downgradient drinking water, surface water, and population areas.
- Description of response actions taken or planned.

7. Important Regulatory Definitions:

Underground storage tank is defined as any tank (including its connected piping) holding an "accumulation of regulated substances" that has 10 percent or more of its volume underground. Federal regulations and most state regulations contain a list of tanks that are exempt from regulation.

---

21 If the release has been reported pursuant to CERCLA, that report will satisfy this requirement.

4.2.5 CWA - Reportable Water Pollutant Discharges

1. Reporting Triggers:

Discharge of oil or hazardous substances in harmful quantities from a vessel or facility into or upon navigable waters of the United States.

2. Reportable Substance:

Oil or hazardous substances, as defined in 40 C.F.R Parts 116 and 117.[22]

3. Reportable Quantity (RQ):

a. Release of oil in "harmful" quantities. Harmful quantities are considered those that violate applicable water quality standards or causes a sheen on nearby surface water.[23]

b. Release of a hazardous substance RQ into navigable waters.[24]

4. Person Responsible for Reporting:

The person in charge of a vessel or facility.

5. Person/Agency to Notify:

Report to the National Response Center (24-hour) at (800) 424-8802 or (202) 267-2675.

6. Timing/Form of Notification:

Immediate notification is required upon *knowledge* of the discharge of a harmful quantity of oil or RQ of a hazardous substance.

If there is a release of an RQ of a hazardous substance into navigable waters, a single report to the National Response Center by the person in charge will satisfy the notification requirements of both the Clean Water Act and CERCLA.

---

22 EPA has established a list of CWA hazardous substances at 40 CFR 116.4. It is important to note that this list of hazardous substances is a subset of the CERCLA list of hazardous substances found at 40 CFR 302.4.

23 40 CFR 110.3.

24 EPA has specified RQs for each designated CWA hazardous substance at 40 CFR 117.3, which should generally be consistent with the RQs for the CERCLA hazardous substances under 40 CFR 302.4.

7. Important Regulatory Definitions:

Oil: Oil is defined as "oil of any kind or in any form, including, but not limited to, petroleum, fuel oil, sludge, oil refuse, and oil mixed with wastes other than dredged spoil."[25]

8. Reporting Exclusions:

a. National Pollutant Discharge Elimination System (NPDES) permitted discharges.[26]

b. Discharges permitted under the International Convention for Prevention of Pollution of the Sea by Oil treaty.[27]

c. Discharges of oil from properly functioning vessel engines.[28]

4.2.6 TSCA - Reportable Chemical Substance Releases

1. Reporting Triggers:

a. *Chemical Substances:* Environmental contamination that could result in substantial risk of injury to health or the environment.

b. *PCBs:* A spill of polychlorinated biphenyls (PCBs) in a reportable quantity.

2. Reportable Substance:

a. Chemical substances and mixtures of them.[29]

b. PCBs.

3. Reportable Quantity (RQ):

a. *Chemical Substances:* Quantity of material that represents a substantial risk of injury to health or the environment.

b. *PCBs:* Spills of PCBs with a concentration of 50 ppm or greater and in a quantity exceeding 10 pounds or in any quantity that directly contaminates surface water, sewers, drinking water supplies, grazing lands, or vegetable gardens.

---

25 CWA 311 (a)(1).

26 CWA 311(a)(2).

27 CWA 311(b)(3)(A).

28 40 CFR 110.7.

29 TSCA 3, 15 USC 2602.

4. Person Responsible for Reporting:

a. *Chemical substances:* Business entities or individual employees.

b. *PCBs:* Owner of the PCB equipment or facility or a designated agent.

5. Person/Agency to Notify:

Notification to the Toxics Section of the EPA Regional Office.

6. Timing/Form of Notification:

a. *Chemical Substances:* Immediately verbal notification upon knowledge of a reportable incident.[30]

A written follow-up report must be submitted to EPA's TSCA office in Washington, D.C. within 15 days of the incident.

The written report must contain the following information:[31]

- Statement that the notice is being submitted in accordance with TSCA Section 8(e)
- Job title, name, address, telephone number, and signature of the person reporting and the facility
- Chemical substance
- Summary of adverse effects, describing the nature and extent of the risk involved
- Source of information.

Note: Reporting is not required under TSCA if the event has been reported to EPA under a mandatory reporting provision of another statute administered by EPA, such as CERCLA.

b. *PCBs:* Oral notification in the shortest possible time after discovery, but in no case later than 24 hours after discovery. No written follow-up notice is required.

4.3 STATE REPORTING AND RECORDKEEPING PROCEDURES

Since most state pollution control laws mirror or expand upon their federal law counterparts, state environmental reporting and recordkeeping requirements largely parallel those contained in the federal environmental laws and regulations. Therefore, the reporting and recordkeeping procedures outlined in Section 4.2 can be followed in most cases. Generally, companies will need to review the reporting and recordkeeping requirements of the state environmental laws for any

---

30 TSCA 6, 15 USC 2607(e).
31 40 CFR 761.125(a)(i).

procedures and reporting information that may differ from the federal requirements.[32] An important difference may be found in the state classification and listing of regulated substances and wastes. For example, state laws may regulate additional hazardous substances and/or the reportable quantity of certain substances may be more stringent than under federal law.

A company's environmental management personnel, regulatory compliance officers, and legal counsel should maintain current copies of state environmental laws and regulations, and contact state environmental authorities for copies of reporting forms and information concerning state reporting procedures. Since it is not possible here to detail the reporting and recordkeeping procedures of each state's environmental laws and regulations, only a survey of several representative states is provided. The state summaries that follow primarily focus on those procedures that differ or place additional reporting demands from the federal reporting and recordkeeping procedures.

4.3.1 California

1. Hazardous Substance Releases:

Generally follow the procedures outlined for CERCLA hazardous substance releases, as outlined in Section 4.2.1.

Emergency Planning and Community Right-to-Know Reporting: Immediately report any release or threatened release of a hazardous substance if there is a reasonable belief that it poses a significant or potential hazard to human health, property, or the environment.[33] Verbally report to the California Office of Emergency Services (24-hour) at (800) 852-7550 or (916) 262-1621 and the local administering agency at 911.

The oral report must include the following information:

- Exact location of the release.
- Name of the person reporting the release.
- Quantity and types of hazardous substances involved.
- Potential hazards posed by the release.

A written follow-up report may be required within 30 days of the incident, which must include the following information:

- Name of the business.
- Name and phone number of a contact person with detailed information about the release.

---

32 See Chapter 2 for a review and summary of the various state environmental law requirements.
33 See Cal. Code of Regs., tit. 19, section 2703.

- Date of incident, time that verbal notification was given, OES control number provided at time of verbal notification.
- Information pertaining to the location of the release.
- Information concerning the specific chemical released (If more than one chemical was released, one report form must be completed for each chemical released).
- Information about the quantity, time, and duration of the release.
- All actions taken to respond to and contain the release.
- Information on any health effects that occurred or might result from the release.
- Information on the type of medical attention required from exposure to the chemical released.

Written reports are to be submitted to the following address:

Chemical Emergency Planning and Response Commission (CEPRC)
Local Emergency Planning Committee (LEPC)
Attn: Section 304 Reports
2800 Meadowview Road
Sacramento, CA 95832

2. Hazardous Waste Discharges:

A hazardous waste generator must provide immediate oral notification of any hazardous waste release, fire, or explosion that could threaten human health or the environment outside the facility to the California Office of Emergency Services (24-hour) at (800) 852-7550 or (916) 262-1621.

The oral report must include:

- Name and telephone number of the facility.
- Person reporting.
- Time and type of incident.
- Name and quantity of hazardous substance involved.
- The extent of injuries, and any potential hazards.

A written follow-up report must be submitted within 15 days to the California Environmental Protection Agency at the following address:

California EPA
Department of Toxic Substances Control
1020 9th Street, Suite 300
Sacramento, CA 95814

The written report must update the information provided in the oral report and provide an estimation of the quantity and disposition of material recovered from the incident.[34]

3. Underground Storage Tank Releases:

Report all unauthorized releases within 24 hours to local officials and the state Office of Emergency Services (24-hour) at (800) 852-7550 or (916) 262-1621.

Within 5 days, a written report of the incident must be submitted to local officials which contains the following information:

- Type, quantity, and concentration of hazardous substances released.
- Results of investigations concerning the extent of contamination.
- Cleanup actions already undertaken or planned, along with costs.
- Method and location of disposal of contaminated substances.
- Proposed method for replacement or repair of primary and secondary containers.
- Facility operator's name and telephone number.

For new underground storage tanks, releases can be reported on regular monitoring reports if they:

- Are from a primary containment system that do not escape from the secondary containment system;
- Do not increase the hazard of fire or explosion;
- Can be cleaned up within 8 hours; and
- Still alow the lead detection system to be reactivated within 8 hours.[35]

4. Water Pollution Discharges:

Report immediately any discharges of hazardous substances exceeding reportable quantities established under CERCLA to the state Office of Emergency Services (24-hour) at (800) 852-7550 or (916) 262-1621.

Report discharges exceeding levels authorized in state water pollutant discharge permits to the nearest Regional Water Quality Control Board. Oral notification of any noncompliance that may endanger health or the environment must be provided within 24 hours from acquiring knowledge of the circumstances.

---

34 See Cal. Code of Regs., tit. 22, section 66265.56.
35 See Cal. Admin. Code. tit. 23, sections 2650, 2652.

A written report must be submitted within 5 days of receiving knowledge of the noncompliance. The written report must contain the following information:

- Description of the noncompliance and its cause.
- Period of noncompliance, including exact dates and times.
- If the noncompliance has not been corrected, the anticipated time it is expected to continue.
- Steps taken or planned to reduce, eliminate, and prevent recurrence of the noncompliance.

Immediate notification to the state Office of Emergency Services (24-hour) at (800) 852-7550 or (916) 262-1621 is required for waste water discharges that exceed 1,000 gallons that result from the diversion of waste water from a collection, treatment, or disposal system.[36]

4.3.2 Connecticut

1. Hazardous Substance Releases:

Verbally report immediately petroleum or chemical substance discharges on land or into waters of the state to the state Department of Environmental Protection at (203) 566-4633 or (203) 566-3338 (24-hour emergency).

A written report must be submitted within 48 hours of the incident, containing the following information:

- Date, time, and person who gave verbal notification of the discharge.
- Time and date of discharge.
- Location of the pollution or contamination.
- Type of petroleum, chemical pollutant, or other contaminant released.
- Quantity of discharge.
- Cause of pollution or contamination.

Emergency Planning and Community Right-to-Know Reporting: Reportable releases (see Section 4.2.2) of EPCRA hazardous substances must be reported immediately to the State Emergency Response Commission at (203) 566-4856 or (203) 566-3338 (24-hour).

2. Hazardous Waste Discharges:

For releases of hazardous wastes that pose a threat to human health outside the facility or if the waste generator knows that a spill has reached surface waters, provide immediate oral notification to the National Response Center (24-hour) at (800) 424-8802 or (202) 267-

---

[36] See Cal. Code of Regs., tit., 23, sections 2250, 2260.

2675 and the state Department of Environmental Protection at (203) 566-4633 or (203) 566-3338 (24-hour emergency).

The oral report should indicate the following:[37]

- Name, address, and EPA Identification Number of the generator.
- Date, time, and type of incident.
- Quantity and type of hazardous waste involved.
- Extent of any injuries.

3. Underground Storage Tank Releases:

Immediately report underground petroleum storage tank releases to the local State Police office.

4. Water Pollution Discharges:

Report within 2 hours (or start of next business day for overnight violations) any discharges exceeding levels authorized in state water pollutant discharge permits which (1) may endanger health or the environment or (2) are greater than 2 times the allowable level for violations set for maximum daily limitations in a discharge permit. Oral notification must be given to the state Department of Environmental Protection, Water Compliance Unit at (203) 566-7139 or (203) 566-3338 (24-hour).

A written report must be submitted within 5 days of receiving knowledge of the noncompliance. The written report must contain the following information:

- Description of the noncompliance and its cause.
- Period of noncompliance, including exact dates and times.
- If the noncompliance has not been corrected, the anticipated time it is expected to continue.
- Steps taken or planned to reduce, eliminate, and prevent recurrence of the noncompliance.

The written report must be submitted to:

Connecticut Department of Environmental Protection
Water Management Bureau
State Office Building
165 Capitol Avenue
Hartford, CT 06106

---

37 See Conn. Gen. Stat. 22a-449(c)-102, incorporating 40 CFR 262.34(d).

4.3.3 Florida

1. Hazardous Substance Releases:

Within one working day, orally report the release of hazardous substances which exceed CERCLA reportable quantities for a 24-hour period (see Section 4.2.1) by calling the Florida Warning Point Number (24-hour), (904) 488-1320.[38]

Emergency Planning and Community Right-to-Know Reporting: Reportable releases (see Section 4.2.2) of EPCRA hazardous substances must be reported immediately to the State Emergency Response Commission at (904) 488-4915 or (904) 488-1320 (24-hour).

A written follow-up report may be required. Written reports are to be submitted to the following address:

    Florida Emergency Response Commission
    Secretary, Florida Department of Community Affairs
    2740 Centerview Drive
    Tallahassee, FL 32399

2. Hazardous Waste Discharges:

For releases of hazardous wastes that pose a threat to human health outside the facility or if the waste generator knows that a spill has reached surface waters, provide immediate oral notification to the National Response Center (24-hour) at (800) 424-8802 or (202) 267-2675 and the Florida Warning Point Number (24-hour), (904) 488-1320.

The oral report should indicate the following:[39]

- Name, address, and EPA Identification Number of the generator.
- Date, time, and type of incident.
- Quantity and type of hazardous waste involved.
- Extent of any injuries.
- Estimated quantity and disposition of any recovered material.

3. Underground Storage Tank Releases:

Within one working day of discovery, report to the state Department of Environmental Regulation:

---

38 See Fla. Admin. Code, tit. 17, section 17-150.300.

39 See Fla. Admin. Code, tit. 17, section 17-730.160, incorporating 40 CFR 262.34(d).

a. Petroleum discharges exceeding 25 gallons on pervious surfaces[40] or that cause a sheen on nearby surface waters.[41]

b. A release of CERCLA hazardous substances which exceeds reportable quantities.[42]

c. Suspected releases confirmed by:

- Released regulated substances or pollutants discovered in the surrounding area;
- Unusual and unexplained storage system operating conditions;
- Monitoring results from a leak detection method or from a tank closure assessment that indicate that a release may have occurred; or
- Manual tank gauging results for tanks of 550 gallons or less, exceeding ten gallons per weekly test or five gallons averaged over four consecutive weekly tests.

Within ten days of the receipt of a test result, report the results of tank tightness testing that exceed allowable tolerances.

4. Water Pollution Discharges:

Report discharges exceeding levels authorized in state water pollutant discharge permits to the the Florida Warning Point Number (24-hour), (904) 488-1320. Oral notification of any noncompliance that may endanger health or the environment must be provided within 24 hours from acquiring knowledge of the noncompliance.

A written report must be submitted to the nearest state Department of Environmental Regulation office within 5 days of receiving knowledge of the noncompliance. The written report must contain the following information:

- Description of the noncompliance and its cause.
- Period of noncompliance, including exact dates and times.
- If the noncompliance has not been corrected, the anticipated time it is expected to continue.
- Steps taken or planned to reduce, eliminate, and prevent recurrence of the noncompliance.

---

40 See Fla. Admin. Code, tit. 17, section 17-761.460.

41 If the petroleum spill or overfill is less than 25 gallons, the release must be reported if it cannot be cleanup up within 24 hours.

42 See Fla. Admin. Code, tit. 17, section 17-761.460(2).

4.3.4 Iowa

1. Hazardous Substance Releases:

Within 6 hours of discovery of a "hazardous condition" involving the release of a hazardous substance, verbally report the incident to the Iowa Department of Natural Resources at (515) 281-8694.

Hazardous Condition is defined as "any situation involving the actual, imminent or probable spillage, leakage, or release of a hazardous substance onto the land, into a water of the state or into the atmosphere which, because of the quantity, strength and toxicity of the hazardous substance, its mobility in the environment and its persistence creates an immediate or potential danger to the public health or safety."

Within 30 days, a written follow-up report must be submitted to:

Iowa Department of Natural Resources
Henry A. Wallace Building
900 East Grand Avenue
Des Moines, IA 50319

The verbal notification and written report should contain the following information:[43]

- Exact location of the hazardous condition.
- Time and date of the occurrence or discovery of the incident.
- Type and quantity of substance released.
- Whether hazardous condition occurred in air, land, or water.
- Name, address, and phone number of the responsible party.
- Time and date of initial verbal notification.
- Weather conditions at the time of the incident.
- Name, address, and phone number of the person reporting the incident.
- Name and phone number of person closest to the scene who can be contacted for further information.
- Information concerning the events leading to the incident.
- Initial measures or corrective action undertaken.

Emergency Planning and Community Right-to-Know Reporting: Reportable releases (See Section 4.2.2) of EPCRA hazardous substances must be reported immediately to the Iowa Department of Natural Resources at (515) 281-8694.

A written follow-up report may be required. Written reports are to be submitted to the following address:

---

43 See Iowa Admin. Code, Division 567, tit. X, ch. 131, section 131.2.

Iowa Department of Natural Resources
Henry A. Wallace Building
900 East Grand Avenue
Des Moines, IA 50319

2. Hazardous Waste Discharges:

TSD Facilities: For generators of 1,000 kg or more of hazardous waste per month, if a release poses a potential threat to human health or the environment outside of the facility, the emergency coordinator must provide immediate verbal notification to the National Response Center at (800) 424-8802 and the Iowa Department of Natural Resources at (515) 281-8694.

The verbal report should provide the following information:[44]

   - Name and phone number of the person reporting.
   - Name and address of the facility.
   - Time and type of incident.
   - Name and quantity of substances involved.
   - Extent of any injuries.
   - Possible hazards to human health and the environment outside of the facility.

Waste Generators: For generators of between 100 and 1,000 kg of hazardous waste per month, if a release could threaten human health outside the facility or the generator knows the spill has reached surface water, immediately porvide verbal notification to the National Response Center at (800) 424-8802 and the Iowa Department of Natural Resources at (515) 281-8694.

The verbal report should provide the following information:[45]

   - Name, address, and EPA Identification Number of the generator.
   - Date, time, and type of incident.
   - Quantity and type of hazardous waste involved.
   - Extent of any injuries.

3. Underground Storage Tanks Releases:

If a "hazardous condition" exists, follow the procedures outlined above under hazardous substances.

Verbally report the following releases within 24 hours of discovery to the state Department of Natural Resources at (515) 281-8694:

---

44 See Iowa Admin. Code, Division 567, tit. X, ch. 141, section 141.3, referring to 40 CFR 265.56.
45 See Iowa Admin. Code, Division 567, tit. X, ch. 141, section 141.3, adopting 40 CFR 262.34(a).

a. Petroleum spills or overfills exceeding 25 gallons[46] or that cause a sheen on nearby surface waters.[47]

b. A release of CERCLA hazardous substances which exceeds reportable quantities.

Suspected releases can be confirmed by:

- Presence of regulated substances in soil, ground water, basements, sewers and utility lines, nearby surface waters, or wells.
- Unusual and unexplained storage system operating conditions, such as sudden loss of a substance from a tank or the unexplained presence of water in a tank.
- Monitoring results from a leak detection method or from a tank closure assessment that indicate that a release may have occurred.

4.3.5 New Jersey

1. Hazardous Substance Releases:

Provide immediately verbal notification of any hazardous substance releases to the New Jersey Department of Environmental Protection at (609) 292-7172 or (609) 882-2000. The verbal notification must be made within 15 minutes[48] and provide the following information:

- Name, title, affliation, address, and phone number of the person reporting the discharge.
- Specific location of the discharge.
- Common name of the hazardous substance discharged.
- Estimate of the quantity of hazardous substance released.
- Date and time the discharge began; date and time the discharge was discovered; and date and time the discharge ended.
- Actions proposed to contain, clean up, and remove the hazardous substances discharged.
- Name and address of any person responsible for the discharge.

Within 30 days, a written discharge confirmation report must be submitted to the following address:

---

[46] See Iowa Admin. Code, Division 567, tit. X, ch. 135, section 135.6.

[47] If the petroleum spill or overfill is less than 25 gallons, the release must be reported if it cannot be cleanup up within 24 hours.

[48] Unless the person responsible for the discharge can show, by clear and convincing evidence, that any delay was reasonable.

New Jersey Department of Environmental Protection
Bureau of Discharge Prevention
Attn: Discharge Confirmation Report
401 East State Street
CN-027
Trenton, NJ 08625-0027

The written report must contain the following information:[49]

- Name, address, and phone number of the person(s) who reported the discharge.
- Name, address, and phone number of the person submitting the written discharge confirmation report.
- If not already identified, the name, address, and phone number of the person subject to the discharge reporting requirements.
- Name, address, and phone number of each person who is in any way responsible for the discharge.
- Name, address, and phone number of each owner and operator of the facility, vessel, or vehicle from which the discharge occurred.
- Source of the discharge.
- Location of the discharge.
- List of all hazardous substances discharged with corresponding common name and Chemical Abstract Service number for each.
- List showing the quantities of each hazardous substance discharged.
- Date and time when the discharge began; date and time when the discharge was discovered; date and time when the discharge ended; and date and time when the agency was notified.
- Detailed description of measures undertaken to contain, clean up, and remedy the discharge, including a summary of costs incurred, and proof of proper disposal of discharged substances.
- Description of corrective actions or countermeasures undertaken, including a description pf equipment repairs or replacements.
- Description of additional preventive measures undertaken or proposed to minimize the possibility of recurrence.
- Name, address, and phone numbers of all entities involved in containment, cleanup, or removal of the discharge.
- Description of the type, quantity, location, and date of all samples taken at or around the site of the discharge, whether before, during, or after any containment, cleanup, or removal.
- The results of the sample analyses.
- For major facilities, a certification of financial responsibility as required under state law.
- Any information needed to supplement that already given to the agency or to correct any false, inaccurate, or misleading information.

---

49 See N.J. Admin. Code, tit. 7, ch. 1E, subchapter 5.

- Certifictions concerning the accuracy and knowledge of the information submitted.

Emergency Planning and Community Right-to-Know Reporting: Reportable releases (See Section 4.2.2) of EPCRA hazardous substances must be reported immediately to the New Jersey Department of Environmental Protection at (609) 292-7172.

A written follow-up report may be required. Written reports are to be submitted to the following address:

New Jersey Department of Environmental Protection ·
Division of Responsible Party Site Remediation
Bureau of Communication and Support Service
CN-411
Trenton, NJ 08625-0411

2. Hazardous Waste Discharges:

New Jersey state hazardous waste reporting procedures mirror those under the federal RCRA program (See Section 4.2.3) and should be performed in conjuction with the more stringent state notification requirements for hazardous substances identified above.

TSD Facilities: If a release poses a potential threat to human health or the environment outside of the facility, the emergency coordinator must provide immediate verbal notification to the National Response Center at (800) 424-8802 and the New Jersey Department of Environmental Protection at (609) 292-7172, as well as local authorities if evacuation of local areas is advisable.

The verbal report should provide the following information:

- Name and phone number of the person reporting.
- Name and address of the facility.
- Time and type of incident.
- Name and quantity of substances involved.
- Extent of any injuries.
- Possible hazards to human health and the environment outside of the facility.
- Proposed actions to respond to the incident.

Within 15 days, a written follow-up report must be submitted to the state Department of Environmental Protection including and updating the information in the verbal report and describing the quantity and disposition of any recovered materials.[50]

---

[50] See N.J. Admin. Code, tit. 7, ch. 26, section 26-9.3.

3. Underground Storage Tanks Releases:[51]

Provide immediate verbal notification of confirmed releases to the state Department of Environmental Protection at (609) 292-7172 and the local Health Agency.[52]

Provide the following information with the verbal notification:

- Type and estimated quantity of the substance released.
- Location of the release.
- Action being taken to contain, clean up, and/or remove the substance.
- Any other information requested by the agency.

Releases can be confirmed by:

- Test, sampling, or monitoring results from a leak or discharge detection method.
- Analysis of ground water samples from the immediate area of the tank system performed by a certified laboratory.
- Results from a closure plan conducted under state law.
- Results from an investigation of a suspected release.
- Any other method, including visual inspection.

If the tank holds hazardous substances, immediately report any release of CERCLA hazardous substances exceeding reportable quantities to the National Response Center at (800) 424-8802.

4. Water Pollution Discharges:

Orally report any discharges exceeding levels authorized in state water pollutant discharge permits which may endanger health or the environment. Oral notification must be given to the state Department of Environmental Protection, Water Compliance Unit at (203) 566-7139 or (203) 566-3338 (24-hour), as follows:

Within 2 hours of acquiring knowledge of the noncompliance provide a verbal report:

- Describing the discharge.
- Explaining the steps being taken to determine the cause of the noncompliance.
- Explaining the steps being taken to reduce and eliminate the noncomplying discharge.

---

51 See N.J. Admin. Code, tit. 7, ch. 14B, section 14B-7.3.

52 Persons required to report releases include the owner or operator of the tank system, or the contractor hired to install, remove, or test the underground storage tank system.

Within 24 hours of acquiring knowledge of the noncompliance provide a verbal report explaining:

- Period of noncompliance, including dates and times, and if not corrected, the anticipated time until the discharge will return to compliance status.
- Cause of the noncompliance.
- Steps being taken to reduce, eliminate, and prevent recurrence.

A written report must be submitted within 5 days of receiving knowledge of the noncompliance. The written report must contain the same information provided in the verbal notification and be submitted to the state Department of Environmental Protection at the following address:

New Jersey Department of Environmental Protection
Assistant Director, Water Quality Management
Division of Water Resources
CN-029
Trenton, NJ 08625-0029

4.3.6 New York

1. Hazardous Substance Releases:

Within 2 hours, orally report the release of a hazardous substance which exceeds reportable quantities to the New York Department of Environmental Conservation's Spill Hotline at (800) 457-7362 or (518) 457-7362.[53]

Emergency Planning and Community Right-to-Know Reporting: Reportable releases (See Section 4.2.2) of EPCRA hazardous substances must be reported immediately to the New York Department of Environmental Conservation's Spill Hotline at (800) 457-7362 or (518) 457-7362.

A written follow-up report may be required. Written reports are to be submitted to the following address:

New York State Emergency Response Commission
Bureau of Spill Prevention and Response
Room 326
50 Wolf Road
Albany, NY 12233-3510

---

53 See N.Y. Codes, Rules & Regs., tit. 6, part 595, section 595.2. A spill or overfill to a secondary containment system does not have to be reported if the spill or overfill is completely contained within 24 hours, there is complete control over the spill or overfill, and the total volume of the substance spilled is recovered or accounted for.

2. Hazardous Waste Discharges:

TSD Facilities: For generators of 1,000 kg or more of hazardous waste per month, if a release poses a potential threat to human health or the environment outside of the facility, the emergency coordinator must provide immediate verbal notification to the National Response Center at (800) 424-8802 and the New York Department of Environmental Conservation's Spill Hotline at (800) 457-7362 or (518) 457-7362.

The verbal report should provide the following information:[54]

- Name and phone number of the person reporting.
- Name and address of the facility.
- Time and type of incident.
- Name and quantity of substances involved.
- Extent of any injuries.
- Possible hazards to human health and the environment outside of the facility.

Waste Generators: For generators of between 100 and 1,000 kg of hazardous waste per month, if a release could threaten human health outside the facility or the generator knows the spill has reached surface water, immediately provide verbal notification to the National Response Center at (800) 424-8802 and the New York Department of Environmental Conservation's Spill Hotline at (800) 457-7362 or (518) 457-7362.

The verbal report should provide the following information:[55]

- Name, address, and EPA Identification Number of the generator.
- Date, time, and type of incident.
- Quantity and type of hazardous waste involved.
- Extent of any injuries.

Within 15 days, a written follow-up report must be submitted which includes and updates the information provided in the verbal notification and which describes the quantity and disposition of any recovered materials. The written report must be submitted to:

New York Department of Environmental Conservation
Commissioner
50 Wolf Road
Albany, NY 12233

---

54 See N.Y. Codes, Rules & Regs., tit. 6, ch. 370, section 370.1(e), referring to 40 CFR 265.56.
55 See N.Y. Codes, Rules & Regs., tit. 6, ch. 370, section 370.1(e), incorporating 40 CFR 262.34(a).

3. Underground Storage Tanks Releases:

Releases from petroleum underground storage tanks are reportable in accordance with the state procedures pertaining to reporting of oil spills.[56] Immediate notification (in no case later than 2 hours) must be given to the state Bureau of Spill Prevention and Response at (800) 457-7362 or (518) 457-7362 regarding:

a. Petroleum spills and discharges into state waters, or onto lands which might flow or drain into state waters, or into waters outside the state which may damage lands, waters, and natural resources within the state.

b. Spills and discharges from any bulk storage tank holding 1,100 gallons of petroleum or other liquid likely to pollute state waters.[57]

The spill notification must provide the following information:

- Name of person making the report and that person's relationship to any person who might be responsible for the discharge.
- Time and date of the discharge.
- Probable source of the discharge.
- Type of pertroleum discharged.
- Possible health or fire hazards.
- Amount of petroleum disharged.
- All actions being taken or to be taken to clean up and remove the discharge.
- Personnel presently at the scene of the discharge.
- Other government agencies that have been or will be notified.

c. Report releases of a Reportable Quantity of a hazardous substance from an underground storage tank in accordance with the procedures for reporting hazardous substance releases explained above.

4. Water Pollution Discharges:

Report within 24 hours any discharges exceeding levels authorized in state water pollutant discharge permits which may endanger health or the environment.[58] Oral notification must be given to the appropriate Regional Office of the New York Department of Environmental Conservation, Division of Water.

---

56 See N.Y. Envtl Conserv. Law 17-1001, 37-0105.

57 A leaking tank also includes failed tank tests performed under the state's bulk storage program.

58 In addition, notification must be given within 24 hours concerning: (1) any unanticipated bypass thath may violate a permit limit; or (2) any upset that may violate a permit limit; or (3) any excursion over the daily maximum limits for any parameter identified in the permit.

A written report must be submitted within 5 days of receiving knowledge of the noncompliance.[59] The written report must contain the following information:

- Description of the noncompliance and its cause.
- Period of noncompliance, including exact dates and times.
- If the noncompliance has not been corrected, the anticipated time it is expected to continue.
- Steps taken or planned to reduce, eliminate, and prevent recurrence of the noncompliance.

The written report must be submitted to:

New York State Department of Environmental Conservation
Division of Water
Operation Chief, Compliance Section
Room 320
50 Wolf Road
Albany, NY 12233-0001

4.3.7 Oregon

1. Hazardous Substance Releases:

Immediately report releases of hazardous substances which exceed reportable quantities for any 24-hour period by calling the Oregon Emergency Response System (24-hour) at (800) 452-0311 or (503) 378-4124.

Emergency Planning and Community Right-to-Know Reporting: Reportable releases (See Section 4.2.2) of EPCRA hazardous substances must be reported immediately to the Oregon Emergency Response System (24-hour) at (800) 452-0311 or (503) 378-6377.

A written follow-up report may be required. Written reports are to be submitted to the following address:

Oregon Emergency Management Division
Emergency Response System
595 Cottage St., N.E.
Salem, OR 97310

2. Hazardous Waste Discharges:

TSD Facilities: For generators of 1,000 kg or more of hazardous waste per month, if a release poses a potential threat to human health or the environment outside of the facility, the emergency coordinator must provide immediate verbal notification to the National Response

---

59 See N.Y. Codes, Rules & Regs., tit. 6, ch. X, part 754, referring to 40 CFR 122.41(l).

Center at (800) 424-8802 and the state Emergency Response System at (800) 452-0311 or (503) 378-4124.

The verbal report should provide the following information:[60]

- Name and phone number of the person reporting.
- Name and address of the facility.
- Time and type of incident.
- Name and quantity of substances involved.
- Extent of any injuries.
- Possible hazards to human health and the environment outside of the facility.

Waste Generators: For generators of between 100 and 1,000 kg of hazardous waste per month, if a release could threaten human health outside the facility or the generator knows the spill has reached surface water, immediately provide verbal notification to the National Response Center at (800) 424-8802 and the state Emergency Response System at (800) 452-0311 or (503) 378-4124.

The verbal report should provide the following information:[61]

- Name, address, and EPA Identification Number of the generator.
- Date, time, and type of incident.
- Quantity and type of hazardous waste involved.
- Extent of any injuries.

Within 15 days, a written follow-up report must be submitted which includes and updates the information provided in the verbal notification and which describes the quantity and disposition of any recovered materials.

The written report must be submitted to:

Oregon Emergency Management Division
Emergency Response System
595 Cottage St., N.E.
Salem, OR 97310

3. Underground Storage Tanks Releases:

Within 24 hours of discovery, the UST owner or operator must report the following releases to the state Emergency Response System at (800) 452-0311:[62]

---

60 See Or. Admin. Rules, ch. 340, section 340-102-010, referring to 40 CFR 265.56.

61 See Or. Admin. Rules, ch. 340, section 340-102-010, incorporating 40 CFR 262.34(a).

62 See Or. Admin. Rules, ch. 340, section 340-150-001(4)(a), incorporating 40 CFR Part 280.

a. Petroleum discharges that exceed 25 gallons or that cause a sheen on nearby surface waters.[63]

b. A release of hazardous substances which exceeds CERCLA reportable quantities.

Suspected releases can be confirmed by:

- Presence of regulated substances in soil, ground water, basements, sewers and utility lines, nearby surface waters, or wells.
- Unusual and unexplained storage system operating conditions, such as sudden loss of a substance from a tank or the unexplained presence of water in a tank.
- Monitoring system results.

4. Water Pollution Discharges:

Within 24 hours, the holder of a pollutant discharge permit shall orally report any excess discharges that may endanger human health or the environment to the state Emergency Response System (24-hour) at (800) 452-0311.[64]

The following additional types of incidents must be reported within 24 hours:

- Any unanticipated bypass or system upset that exceeds permit limitations.
- Any violation of a maximum daily discharge limitation for which the state requires 24-hour reporting in the permit.

Within 5 days, a written report must be submitted that provides the following information:

- Description of the noncompliance and its cause.
- Period of the discharge, including times and dates.
- If uncorrected, an estimate of how long the discharge is expected to continue.
- Steps taken to reduce, eliminate, and prevent recurrence of the problem.

The written report must be submitted to:

---

63 If the petroleum spill or overfill is less than 25 gallons, the release must be reported if it cannot be cleanup up within 24 hours.

64 See Or. Admin. Rules, ch. 340, section 340-45-065, referring to the federal NPDES reporting requirements at 40 CFR 122.41(l).

Oregon Emergency Management Division
Emergency Response System
595 Cottage St., N.E.
Salem, OR 97310

4.3.8 Utah

1. Hazardous Substance Releases:

Generally follow the procedures outlined for CERCLA hazardous substance releases, as outlined in Section 4.2.1.

Emergency Planning and Community Right-to-Know Reporting: Reportable releases (see Section 4.2.2) of EPCRA hazardous substances must be reported immediately to the Utah Hazardous Chemical Emergency Response Commission at (801) 536-4100 (8 to 5) or (801) 536-4123 (24-hour).

A written follow-up report may be required. Written reports are to be submitted to the following address:

Utah Hazardous Chemical Emergency Response Commission
Utah Department of Environmental Quality
168 North, 1950 West
Salt Lake City, UT 84114-4840

2. Hazardous Waste Discharges:[65]

TSD Facilities: Within 24 hours, facilities permitted to treat, store, or dispose of hazardous wastes must provide verbal notification of any permit noncompliance which poses a potential threat to human health or the environment outside of the facility. The emergency coordinator must notify the Utah Solid and Hazardous Waste Control Board at (801) 536-4100 (8 to 5) or (801) 536-4123 (24-hour).

The verbal report must provide the following information:

- Information concerning threat to public drinking water or threat to human health or the environment outside the facility.
- Name, address, and telephone number of owner or operator.
- Name, address, and telephone number of the facility.
- Name and quantity of materials involved.
- Extent of any injuries.
- Assessment of the actual or potential hazards to the environment and human health outside the facility
- Estimated quantity and disposition of recovered material.

---

65 See Utah Code Ann. 19-6-101 to 122. "Hazardous waste" is defined at Utah Admin. Code R-315-2-1.

A written follow-up notice is required within 5 days. The following information is required in the written report:[66]

- Description of noncompliance, including exact dates and times.
- If noncompliance is not corrected, anticipated time it is expected to continue.
- Steps taken or planned to reduce, eliminate, and prevent recurrence of noncompliance.

The written report must be submitted to:

Utah Department of Environmental Quality
Division of Environmental Response and Remediation
168 North, 1950 West
Salt Lake City, UT 84114-4840

Hazardous Waste Tank Systems: Facilities permitted by the state to store or treat hazardous waste using tank systems are required to notify the Utah Solid and Hazardous Waste Control Board within 24 hours of the detection of a release of more than one pound of hazardous waste to the environment (or one pound or less if not immediately contained and cleaned up).[67]

A written follow-up report is required within 30 days, which provides the following information:

- Likely route of migration of the release.
- Characteristics of the surrounding soil.
- Results of any monitoring or sampling conducted.[68]
- Proximity to downgradient drinking water, surface water, and populated areas.
- Description of response actions taken or planned.

The written report must be submitted to:

Utah Department of Environmental Quality
Division of Environmental Response and Remediation
168 North, 1950 West
Salt Lake City, UT 84114-4840

Waste Generators: Generators of between 100 and 1,000 kg of hazardous waste per month must immediately report the following types of releases to the Utah Department of Environmental Quality at (801) 536-4100 (8 to 5) or (801) 536-4123 (24-hour).

---

66 See Utah Admin. Code R-315-3-4(6).

67 Notice is not required if the release has been reported under CERCLA.

68 If not available within 30 days, these data must be submitted as soon as they are available.

- A spill of 1 kg of acute hazardous waste.[69]
- A spill of 100 kg of non-acute hazardous waste.
- A spill of less than 100 kg of non-acute hazardous waste which could threaten human health or the environment.

The verbal report should provide the following information:

- Name, telephone number, and address of the person responsible for the spill.
- Name, title, and telephone number of the individual reporting.
- Time and date of the spill.
- Location of the spill.
- Description and amount of the spill.
- Causes of the spill.
- Emergency action taken.

Within 15 days, a written follow-up report must be submitted to the Utah Solid and Hazardous Waste Control Board which includes the following information:[70]

- Person's name, telephone number, and address.
- Date, time, location, and nature of the incident.
- Name and quantity of material involved.
- Extent of any injuries.
- Assessment of actual or potential hazards to human health or the environment.
- Estimated quantity and disposition of recovered material.

The written report must be submitted to:

Utah Department of Environmental Quality
Division of Environmental Response and Remediation
168 North, 1950 West
Salt Lake City, UT 84114-4840

3. Underground Storage Tanks Releases:[71]

Within 24 hours, the UST owner or operator must report the following releases to the Utah Solid and Hazardous Waste Control Board at (801) 536-4100 (8 to 5) or (801) 536-4123 (24-hour).[72]

a. Suspected releases of regulated substances.[73] Suspected releases must be investigated and confirmed within 7 days by:

---

69 "Acute hazardous wastes" are listed at Utah Admin. Code R-315-2.1.9(e).

70 See Utah Admin. Code, R-315-8-10, incorporating 40 CFR 2642.196(d).

71 See Utah Code Ann. 19-6-419, 19-6-423.

72 See Utah Admin. Code, R-451-202-1, incorporating 40 CFR Part 280.

73 "Regulated substances" is defined to include hazardous substances under CERCLA and petroleum. Utah Admin. Code R-311-202-1, incorporating 40 CFR Part 280.

- Presence of regulated substances in soil, groundwater, basements, sewers and utility lines, nearby surface waters, or wells.
- Unusual and unexplained storage system operating conditions, such as sudden loss of a substance from a tank or the unexplained presence of ater in a tank.
- Monitoring system results.

b. Petroleum discharges exceeding 25 gallons or that cause a sheen on nearby surface waters.[74]

c. A release of hazardous substances which exceeds CERCLA reportable quantities.

4. Water Pollution Discharges:[75]

Within 24 hours, the holder of a state water pollution discharge permit shall orally report any excess discharges that may endanger human health or the environment to the Water Quality Board at (801) 536-4100 (8 to 5) or (801) 536-4123 (24-hour).

The following additional types of incidents must be reported within 24 hours:

- Any unanticipated bypass or system upset that exceeds permit limitations.
- Any violation of a maximum daily discharge limitation for which the state requires 24-hour reporting in the permit.

Within 5 days, a written report must be submitted that provides the following information:[76]

- Description of the noncompliance and its cause.
- Period of noncompliance, including exact dates and times.
- If the noncompliance has not been corrected, the anticipated time it is expected to continue.
- Steps taken or planned to reduce, eliminate, and prevent recurrence of the noncompliance.

The written report must be submitted to:

Utah Department of Environmental Quality
Division of Environmental Response and Remediation
168 North, 1950 West
Salt Lake City, UT 84114-4840

---

74 If the petroleum spill or overfill is less than 25 gallons, the release must be reported if it cannot be cleaned up within 24 hours.

75 See Utah Code Ann. 19-5-101 to 119.

76 Utah Admin. Code R-317-8.4.1(12)(f).

Chapter 5

Emergency Response / Waste Minimization / Compliance Audits

5.1   INTRODUCTION

Regardless of a company's size, every company that handles hazardous materials or wastes in ongoing business operations should have an emergency response (or contingency) plan in place. The risk of pollution liability and regulatory noncompliance penalties provide strong incentives for implementing a contingency plan at the company site. Even the remote possibility of a hazardous substance incident should not be taken lightly. Preventative hazardous waste management measures, such as waste minimization,[1] can reduce the potential risk of a release or spill, although accidents are unfortunately bound to occur. A contingency plan is needed to ensure rapid and effective company response to hazardous substance emergencies. The nature of the business, degree of involvement with hazardous materials, and volume and toxicity of hazardous substances handled at the company site will determine how extensive the contingency plan should be.

Environmental laws and regulations may mandate specific content requirements for contingency plans. The Clean Water Act (CWA), Occupational Health and Safety Act, the Hazardous Materials Transportation Act (HMTA), Comprehensive Environmental Response Compensation and Liability Act (CERCLA), and Resource Conservation and Recovery Act (RCRA) all contain provisions regarding emergency response. By way of illustration, the focus here is on the detailed RCRA regulations governing hazardous waste contingency plans.

5.2   HAZARDOUS WASTE CONTINGENCY PLANS

Although under RCRA, only treatment, storage, and disposal (TSD) facilities are required to have a contingency plan in place to handle hazardous waste spills or releases, the mandatory response plan serves as a good model for all companies that manufacture, handle, or generate hazardous substances. The company's environmental manager should review the RCRA contingency plan regulations when designing an appropriate emergency response plan.[2]

5.2.1 RCRA Contingency Plan Components

A "contingency plan" is a document that sets forth an organized, planned, and coordinated course of action to be followed in the event of a fire, explosion, or release of hazardous substances which could threaten human health and safety or the environment. Company personnel should be advised on what to do in the event of an emergency. The contingency plan should be placed in a binder in a location easily accessible to all employees. The RCRA contingency plan must be a written document and contain the following components:

---

1 See 5.4.

2 The RCRA regulations are found at 40 CFR Sections 264.30 - 264.56 and 265.30 - 265.56.

1. Description of actions to be taken by company personnel in the event of a fire, explosion, sudden or nonsudden release;
2. Description of police, fire, hospital, and local emergency response arrangements;
3. Listing of names, addresses, and telephone numbers of all persons qualified to act as emergency coordinators;
4. Listing of all emergency equipment, including location, physical description, and an outline of capabilities; and
5. An evacuation plan for company personnel, including routes and alternate routes.

If the company already has a Spill Prevention, Control, and Countermeasures (SPCC) plan pursuant to the Clean Water Act, the company only needs to amend the plan to incorporate RCRA hazardous waste management components.

In addition to having a copy available on-site, copies of the contingency plan must be filed with the police, fire, hospital, and local emergency response teams. If the plan fails in an emergency, or changes occur regarding the emergency coordinator, the equipment listed, or the facility operations, the contingency plan must be amended to take these changes into account.

5.2.2 Emergency Equipment

Depending on the degree and type of hazards posed at the company site, various types of equipment will be needed to properly monitor operations and protect employees from exposure to hazardous substances.[3] Equipment should be periodically checked and maintained to assure its proper working operation in time of emergency. Equipment should not be stored near hazardous substances and should be easily accessible and available at all times. Consumed items must be replaced and reusable items cleaned or decontaminated. Appropriate equipment for handling a hazardous substance emergency includes:

* Portable fire extinguishers;
* 85 gallon salvage drums (to hold leaking 55 gallon drums);
* Nonreactive gloves - depending on the substance - such as PVC for most corrosive substances, neoprene for petroleum products, and fluoroelastomer for chlorinated solvents;
* Nonreactive absorbent material to absorb spilled substances on floors or around drums, such as vermiculite;
* Nonreactive apron or coveralls selected according to the substance spilled or being handled;[4]
* Transfer pumps to remove spilled substances; and
* Other fire protection, spill control, and decontamination equipment as needs dictate.

---

3 OSHA provides additional worker safety requirements that are beyond the scope of this discussion.
4 Tyvek is a common material.

It also may be wise to have air purifying respirators on hand; however, proper training is required for respirators. OSHA regulations govern the use, type of protection they provide, and the atmospheres suitable for respirators. Training must be performed by a qualified person who is knowledgeable about the OSHA regulations. Other equipment may also be appropriate for the company. Consultation with a qualified safety engineer or industrial hygienist can determine additional equipment needs at the company.

5.2.3 Communication System

In addition to testing and maintenance of appropriate equipment, the company should also have some form of internal communication system to alert and provide instruction to company personnel in case of an emergency. A telephone or hand-held two-way radio should also be immediately accessible to summon external emergency assistance. Further emergency preparedness measures would include maintenance of necessary aisle space to allow unobstructed and rapid movement of personnel, fire protection equipment, spill control equipment, and decontamination equipment to any area of the facility, and maintenance of adequate water pressure and volume to supply water hoses, foam-producing equipment, or water sprinkler systems.

5.2.4 Emergency Coordinator

Every emergency response plan must include designation and training of an emergency coordinator. The RCRA contingency plan requires appointment of an emergency coordinator who is on-site or on call at all times. The emergency coordinator must be familiar with all aspects of the company operation and emergency procedures, and must have the authority to implement the contingency plan. In the event of an emergency, the emergency coordinator must identify the character, source, and extent of the release, and assess any potential hazard to human health and the environment. The coordinator must immediately notify appropriate local authorities when a hazard exists, and promptly arrange for treatment, storage or disposal of recovered waste, contaminated soil, or surface water. A written report of the incident must be reported to EPA within fifteen days. Details of the release should be kept in a log and reported on the annual Toxic Chemical Release Inventory (Form R).

5.2.5 Implementing a Contingency Plan

The primary focus of the contingency plan should be on implementing emergency response procedures tailored to the company's needs, adequately covering the risk to personnel and the environment, and complying with all government regulations. By following the guidelines outlined here, referring to the RCRA contingency plan, and consulting with environmental managers and safety engineers, the company can develop a strong emergency preparedness and emergency response plan.

Other sources worth consulting include EPA's guidance document, *Chemical Emergency Preparedness Program* and DOT's *1990 Emergency Response Guide*, as well as information available from various associations, such as the Chemical Manufacturers Association's *Chemical Awareness and Emergency Response Program*.

Once the company's contingency plan is in place, periodic drills should be conducted to test the effectiveness of equipment, communication systems, evacuation, and overall response to a hypothetical emergency. If possible, involve local authorities when conducting drills.

5.3 POLLUTION PREVENTION

By now, most companies that generate hazardous wastes are aware of the environmental regulatory trend toward pollution prevention. Following many years of relying on various "command-and-control" measures to regulate pollution, Congress, the U.S. EPA, and the states have been incorporating new pollution prevention requirements into the federal environmental law framework. Until recently, the predominant waste management practice had been "end of pipe" treatment or land disposal of hazardous and nonhazardous wastes. While this approach has provided substantial progress in improving the quality of the environment, obvious limits exist on how much environmental improvement can be achieved using methods that manage pollutants after they have been generated.

Pollution prevention is very important as an environmental management tool. It provides a means of reducing wastes that could give rise to environmental liabilities. By implementing waste minimization measures, a company reduces the volume and toxicity of wastes normally generated by a given chemical manufacturing process or business operation. Waste minimization results in economic benefits, competitive advantages, enhanced public image, and reduced environmental risks. Obviously, by having to manage fewer wastes, a company has fewer wastes to dispose of, fewer wastes to report, and reduced risk of spills and releases that contaminate the environment (and less money spent on cleanup and environmental penalties).

5.3.1 Pollution Prevention Act of 1990

In 1990, Congress passed the Pollution Prevention Act of 1990 (PPA)[5] to focus more attention on reducing the volume and toxicity of wastes at the source. Thus, source reduction, recycling, and other waste minimization strategies are fast becoming a significant environmental regulatory compliance issue. In Section 6602(b) of this law,[6] Congress stated a national policy goal to prevent or reduce pollution at the

---

5 Pub. L. 101-508, 42 USC 13101 et seq.

6 42 USC 13101(b).

source whenever feasible. The PPA sets forth a hierarchy of management options in descending order of preference: prevention, environmentally sound recycling, environmentally sound treatment, and environmentally sound disposal.

This statute is basically an enabling act which states congressional commitment to waste reduction and recycling activities and which mandates that U.S. EPA implement pollution prevention strategies and regulations. The PPA also requires that U.S. EPA provide grants to the states to implement their own pollution prevention programs, and requires that U.S. EPA set up an information clearinghouse and conduct pollution prevention research. The PPA also requires that companies report their pollution prevention practices under SARA Title III.

5.3.2 EPA's Pollution Prevention Strategy

Under the PPA, EPA must develop a pollution prevention strategy that reduces pollution at the source. In February 1991, EPA published its pollution prevention strategy, commonly referred to as the Industrial Toxics Project or the "33/50" Initiative.[7] Under the plan, EPA is seeking to reduce releases and off-site transfers of seventeen high-volume EPCRA Section 313 toxic chemicals. These industrial chemicals include known and potential carcinogens, developmental toxins, chemicals that bioaccumulate, ozone-depleting chemicals, and chemicals contributing to ozone at ground level. EPA set a goal to reduce the releases of these chemicals by 33% by the end of 1992, and 50% by the end of 1995.

Under the program, EPA encourages companies to reach the 33 and 50% goals by using the pollution prevention hierarchy. Since the PPA accords source reduction the highest value, EPA correspondingly makes source reduction the preferred method of pollution prevention. Treatment is considered the next best method, and disposal is the least preferred method of pollution prevention. Although participation in the Industrial Toxics Project is completely voluntary and the program goals are not enforceable, companies should take steps to implement pollution prevention techniques because such measures are destined to become mandatory components of pollution control laws.

EPA is implementing guidance on pollution prevention for companies to follow. So far, EPA has released a number of guidance documents, has conducted numerous studies, and is beginning to develop different regulations with pollution prevention requirements. In addition, the states are implementing statutes with various pollution prevention requirements. Fifteen states have statutes that make pollution prevention mandatory, and twelve additional states have statutes with voluntary pollution prevention provisions.

---

7 56 Fed. Reg. 7849 (Feb. 26, 1991).

5.3.3 Hazardous and Solid Waste Amendments to RCRA

Further, with the passage of the Hazardous and Solid Waste Amendments to RCRA in 1984, Congress established a significant new policy concerning hazardous waste management. Specifically, Congress declared that the reduction or elimination of hazardous waste generation at the source should take priority over the management of hazardous wastes after they are generated. In particular, in RCRA Section 1003(b),[8] Congress declared it to be national policy that, whenever feasible, the generation of hazardous waste is to be reduced or eliminated as expeditiously as possible. Waste that is nevertheless generated should be treated, stored, or disposed of so as to minimize the present and future threat to human health and the environment.

In furtherance of the national policy toward pollution prevention, in 1989, EPA published a proposed policy statement on source reduction and recycling.[9] With this policy, EPA announced a commitment to a preventive strategy for reducing or eliminating the generation of pollutants that may be released into the air, land, surface water, or ground water. EPA proposed to incorporate this preventive strategy into its overall mission to protect human health and the environment by making source reduction a priority for every aspect of agency decision-making and planning, with environmentally sound recycling as a second and higher priority over treatment and disposal.

5.4 HAZARDOUS WASTE MINIMIZATION

On May 28, 1993, EPA issued interim final guidance to assist hazardous waste generators and owners and operators of hazardous waste treatment, storage, or disposal (TSD) facilities to comply with the waste minimization certification requirements of RCRA Sections 3002(b) and 3005(h), as amended by the HSWA.[10] Under RCRA Section 3002(b), hazardous waste generators who transport their wastes off-site must certify on their hazardous waste manifests that they have programs in place to reduce the volume or quantity and toxicity of hazardous waste generated to the extent economically practicable. Certification of a waste minimization "program in place" is also required as a condition of any permit issued under section 3005(h) for the treatment, storage, or disposal of hazardous waste at facilities that generate and manage hazardous wastes on-site.

The guidance document fulfills a commitment made by EPA in its 1986 report to Congress entitled *The Minimization of Hazardous Waste*[11] to

---

8 42 USC 6902(b).

9 54 Fed. Reg. 3845 (Jan. 26, 1989).

10 "Guidance to Hazardous Waste Generators on the Elements of a Waste Minimization Program," 58 Fed. Reg. 31114 (May 28, 1993).

11 EPA/530-SW-86-033 (Oct. 1986).

provide additional information to generators on the meaning of the certification requirements added by the HSWA.[12]

5.4.1 What Constitutes "Waste Minimization"?

EPA considers waste minimization, the term employed by Congress in the RCRA statute, to include:

1. Source reduction; and

2. Environmentally sound recycling.

The first category, source reduction, is defined in Section 6603(5)(A) of the Pollution Prevention Act,[13] as any practice which:

1. Reduces the amount of any hazardous substance, pollutant, or contaminant entering any waste stream or otherwise released into the environment (including fugitive emissions) prior to recycling, treatment, or disposal; and

2. Reduces the hazards to public health and the environment associated with the release of such substances, pollutants, or contaminants.

The term includes equipment or technology modifications, process or procedure modifications, reformulation or redesign of products, substitution of raw materials, and improvements in housekeeping, maintenance, training, or inventory control. EPA relies on this definition for use in identifying opportunities for source reduction under RCRA.

The second category, environmentally sound recycling, is the next preferred alternative for managing those pollutants that cannot be reduced at the source. In the context of hazardous waste management, there are certain practices or activities that the RCRA regulations define as "recycling." The definitions for materials that are "recycled" are found in 40 C.F.R. Section 261.1(c).

EPA considers recycling activities that closely resemble conventional waste management activities not to constitute waste minimization. Unfortunately, it is not always easy to distinguish recycling from conventional treatment.[14] Treatment for the purposes of destruction or disposal is not part of waste minimization, but is, rather, an activity that occurs after the opportunities for waste minimization have been pursued.

---

12 See 51 Fed. Reg. 44683 (December 11, 1986).

13 42 USC 13102(5)(a).

14 See 56 Fed. Reg. 7143 (Feb. 21, 1991); 53 Fed. Reg. 522 (Jan. 8, 1988).

Transfer of hazardous constituents from one environmental medium to another also does not constitute waste minimization. For example, the use of an air stripper to evaporate volatile organic constituents from an aqueous waste only shifts the contaminant from water to air. Furthermore, concentration activities conducted solely for reducing volume does not constitute waste minimization unless, for example, concentration of the waste is an integral setup in the recovery of useful constituents prior to treatment and disposal. Similarly, dilution as a means of toxicity reduction would not be considered waste minimization, unless dilution is a necessary step in a recovery or a recycling operation.

5.4.2 Elements of a Hazardous Waste Minimization Program

EPA's guidance on the elements of a waste minimization program is intended to assist companies and individuals to properly certify that they have implemented a program to reduce the volume and toxicity of hazardous waste to the extent "economically practicable." The guidance is directly applicable to generators who generate 1000 or more kilograms per month of hazardous waste ("large quantity" generators) or to owners and operators of hazardous waste treatment, storage, or disposal facilities who manage their own hazardous waste on-site.

Small quantity generators who generate greater than 100 kilograms but less than 1000 kilograms of hazardous waste per month are not subject to the same "program in place" certification requirement as large quantity generators. Instead, they must certify on their hazardous waste manifests that they have "made a good faith effort to minimize" their waste generation. Nevertheless, EPA encourages small quantity generators to develop their own waste minimization programs to show good faith efforts.

. According to the EPA guidance (which is not a formal regulation, and therefore, not enforceable) the following basic elements should be part of most waste minimization programs:

1. Top management support;
2. Characterization of waste generation and waste management costs;
3. Periodic waste minimization assessments;
4. Appropriate cost allocation;
5. Encouragement of technology transfer; and
6. Program implementation and evaluation.

Thus, generators should consider these elements when designing multimedia pollution prevention programs directed at preventing or reducing wastes, substances, discharges, and/or emissions to all environmental media - air, land, surface water, and ground water. Each of these elements is discussed below.

Top Management Support

Top management should support a company-wide effort. There are many ways to accomplish this goal. Some of the methods described below may be suitable for some companies, while not for others. However, some combination of these techniques or similar ones will demonstrate top management support:

1. Make waste minimization a part of the company policy. Put this policy in writing and distribute it to all departments and individuals. Each individual, regardless of status or rank, should be encouraged to identify opportunities to reduce waste generation. Encourage workers to adopt the policy in day-to-day operations and encourage new ideas at meetings and other organizational functions. Waste minimization, especially when incorporated into company policy, should be a process of continuous improvement. Ideally, a waste minimization program should become an integral part of the company's strategic plan to increase productivity and quality.

2. Set explicit goals for reducing the volume and toxicity of waste streams that are achievable within a reasonable time frame. These goals may be quantitative or qualitative. Both can be successful.

3. Commit to implementing recommendations identified through assessments, evaluations, and waste minimization teams.

4. Designate a waste minimization coordinator who is responsible for facilitating effective implementation, monitoring, and evaluation of the program. In some cases (particularly in large multifacility organizations), an organizational waste minimization coordinator may be needed in addition to facility coordinators. In other cases, a single coordinator may have responsibility for more than one facility. In these cases, the coordinator should be involved or be aware of operations and should be capable of facilitating new ideas at each facility. It is also useful to set up self-managing waste minimization teams chosen from a broad spectrum of operations: engineering, management, research and development, sales and marketing, accounting, purchasing, maintenance, and environmental staff personnel. These teams can be used to identify, evaluate, and implement waste minimization opportunities.

5. Publicize success stories. Set up an environment and select a forum where creative ideas can be heard and tried. These techniques can inspire additional ideas.

6. Recognize individual and collective accomplishments. Reward employees that identify cost-effective waste minimization

opportunities. These rewards can take the form of collective and/or individual monetary or other incentives for improved productivity/waste minimization.

7. Train employees on the waste-generating impacts that result from the way they conduct their work procedures. For example, purchasing and operations departments could develop a plan to purchase raw materials with less toxic impurities or return leftover materials to vendors. This approach can include all departments, such as those in research and development, capital planning, purchasing, production operations, process engineering, sales and marketing, and maintenance.

Characterization of Waste Generation and Management Costs

Maintain a waste accounting system to track the types and amounts of wastes as well as the types and amounts of the hazardous constituents in wastes, including the rates and dates they are generated. Each organization must decide the best method to obtain the necessary information to characterize waste generation. Many organizations track their waste production by a variety of means and then normalize the results to account for variations in production rates.

In addition, a waste generator should determine the true costs associated with waste management and cleanup, including the costs of regulatory oversight compliance, paperwork and reporting requirements, loss of production potential, costs of materials found in the waste stream (perhaps based on the purchase price of those materials), transportation/treatment/storage/disposal costs, employee exposure and health care, liability insurance, and possible future RCRA or Superfund corrective action costs. Both volume and toxicities of generated hazardous waste should be taken into account. Substantial uncertainty in calculating many of these costs, especially future liability, may exist. Therefore, each organization should find the best method to account for the true costs of waste management and cleanup.

Periodic Waste Minimization Assessments

Different and equally valid methods exist by which a waste minimization assessment can be performed. Some organizations identify sources of waste by tracking materials that eventually wind up as waste, from point of receipt to the point at which they become a waste. Other organizations perform mass balance calculations to determine inputs and outputs from processes and/or facilities. Larger organizations may find it useful to establish a team of independent experts outside the organization structure, while some organizations may choose teams comprised of in-house experts. Most successful waste minimization assessments have common elements that identify sources of waste and calculate the true costs of waste generation and management. Each organization should decide the best method to use in performing a waste minimization assessment that addresses these two general elements:

1. Identify opportunities at all points in a process where materials can be prevented from becoming a waste (for example, by using less material, recycling materials in the process, finding substitutes that are less toxic and/or more easily biodegraded, or making equipment/process changes). Individual processes or facilities should be reviewed periodically. In some cases, performing complete facility material balances can be helpful.

2. Analyze waste minimization opportunities based on the true costs associated with waste management and cleanup. Analyzing the cost effectiveness of each option is an important factor to consider, especially when the true costs of treatment, storage, and disposal are considered.

Cost Allocation System

If practical and implementable, organizations should appropriately allocate the true costs of waste management to the activities responsible for generating the waste in the first place (e.g., identifying specific operations that generate the waste, rather than charging the waste management costs to "overhead"). Cost allocation can properly highlight the parts of the organization where the greatest opportunities for waste minimization exist; without allocating costs, waste minimization opportunities can be obscured by accounting practices that do not clearly identify the activities generating the hazardous wastes.

Encourage Technology Transfer

Many useful and equally valid techniques have been evaluated and documented that are useful in a waste minimization program. It is important to seek or exchange technical information on waste minimization from other parts of the organization, from other companies, trade associations, professional consultants, and university or government technical assistance programs. EPA- and/or State-funded technical assistance programs (e.g., Minnesota Technical Assistance Program - MnTAP, California Waste Minimization Clearinghouse, EPA Pollution Prevention Information Clearinghouse) are becoming increasingly available to assist in finding waste minimization options and technologies.

Program Implementation and Evaluation

Implement recommendations identified by the assessment process, evaluations, and waste minimization teams. Conduct a periodic review of program effectiveness. Use these reviews to provide feedback and identify potential areas for improvement.

5.4.3 Further Information on Waste Minimization Programs

EPA and the States have worked cooperatively to put in place a variety of technical information and assistance programs that make information on source reduction and recycling techniques available directly to industry and the public. EPA has developed information sources that can be used to provide information directly to industry or through state technical assistance programs. EPA maintains a Pollution Prevention Information Clearinghouse (PPIC), which is a reference and referral source for technical, policy, program, legislative, and financial information on pollution prevention. PPIC's telephone number is (202) 260-1023; the facsimile number is (202) 260-0178. EPA also produces videos and literature on waste minimization that are available to the public, as well as a pollution prevention newsletter.[15]

Examples of general documents that assist organizations with more detailed guidance on conducting waste minimization assessments and developing pollution prevention programs are:

1. *Waste Minimization Opportunity Assessment Manual*, EPA 625/7-88/003, July 1988 (Pub. No. PB 92-216 985), available by calling NTIS at (703) 487-4650;

2. *Facility Pollution Prevention Guide*, EPA/600/R-92/088, available by calling the CERI Publications Unit at EPA's Cincinnati office at (513) 569-7562;

3. *Waste Minimization: Environmental Quality with Economic Benefits*, EPA/530-SW-90-044, April 1990, available by calling the RCRA Information Center at (202) 260-9327.

EPA has also developed numerous waste minimization and pollution prevention documents that are tailored to specific manufacturing and other types of processes, and periodically sponsors pollution prevention workshops and conferences.

5.4.4 Benefits of Waste Minimization

Waste minimization provides additional environmental improvements over "end of pipe" control practices, often with the added benefit of cost savings to generators of hazardous waste and reduced levels of treatment, storage, and disposal. Waste minimization has already been shown to result in significant benefits for industry, as evidenced in numerous success stories documented in available literature.

---

15 To be added to the mailing list, write to: Pollution Prevention News, U.S. EPA, PM-222B, 401 M St., S.W., Washington, DC 20460.

The benefits to companies that pursue waste minimization often include:

1. Minimizing quantities of hazardous waste generated, thereby reducing waste management and compliance costs and improving the protection of human health and the environment;
2. Reducing or eliminating inventories and possible releases of "hazardous chemicals";
3. Possible decrease in future Superfund and RCRA liabilities, as well as future toxic tort liabilities;
4. Improving facility mass/energy efficiency and product yields;
5. Reducing worker exposure; and
6. Enhancing organizational reputation and image.

Waste minimization programs are being implemented by a wide array of organizations. Numerous state governments have also enacted legislation requiring facility specific waste minimization programs, and other states have legislation pending that may mandate some type of facility-specific waste minimization program.

5.5   ENVIRONMENTAL COMPLIANCE AUDITS

Every company needs to protect itself against costly environmental liabilities and penalties. Implementation of safe methods for handling hazardous materials and wastes in day-to-day business operations will minimize the risk of liability. Compliance with the multitude of environmental laws and regulations will minimize the imposition of government penalties. Routine checks of hazardous activities and pollution-generating operations are one way to confirm continued compliance with environmental laws, and continued maintenance of safe business practices at the company site. Periodic environmental audits have become the standard means of assessing site contamination and regulatory compliance.

5.5.1 Environmental Audits

Environmental audits are most frequently undertaken when property is being transferred. Lenders usually require an environmental audit before giving a mortgage, and today only a foolhardy buyer would purchase property without the benefit of a due diligence investigation of the site. Some states, such as New Jersey, Illinois, and Connecticut, require an environmental audit prior to the transfer of property.

An environmental consulting firm is usually hired to carry out the audit. The extent and scope of the audit will vary depending on the particular circumstances. Generally, a Phase I site investigation is undertaken to discover potential contamination or noncompliance with environmental laws and regulations. If actual or potential contamination is found, a more extensive Phase II study (including groundwater and soil sampling) may be required to verify and determine the extent of any environmental problems.

5.5.2 Regulatory Compliance Audits

A separate and distinct component of environmental audits is the regulatory compliance audit. Although a full audit should be implemented whenever property transfers take place, the compliance component is particularly beneficial to assess whether all or selected levels of a company's business operations are in compliance with environmental regulations, internal policies, and accepted business practices. Unlike the due diligence audit, which is most concerned with contamination, the regulatory compliance audit focuses on making sure that company operations meet regulatory standards.

Compliance with laws and regulations is probably the most common driver of environmental audits. Existing laws and regulations cover different types of activities and environmental media. Virtually all of the federal laws and regulations have state and local counterparts. The hallmark of state and local regulations is that they may be "no less stringent" than their federal counterparts. As a result, an effective audit must recognize that compliance with federal law alone will not minimize the risk associated with state and local regulatory compliance.

An environmental consultant can be retained solely for the purpose of conducting a regulatory compliance audit. A full compliance audit can run anywhere from $5,000 to $25,000 per site, depending on the size and nature of the business. A reputable consultant must be chosen who has expertise in completing such an assessment.

5.5.3 Auditing of Company Policies; Environmental Management Systems

Review of company policies should also be included in an environmental audit. Company policies are generally established centrally, but implemented locally. Therefore, it behooves corporate leadership to ensure that an auditing system is in place to maintain environmental standards throughout the company's operations.

Environmental management systems present a unique opportunity for auditing. Most sizable companies have sophisticated management systems for their operations. These may include financial, inventory, and personnel management systems. Many of these systems are now beginning to include environmental management components. One of the benefits of an environmental audit is that it can test the effectiveness of an environmental management system by assessing whether it is operating properly, and whether the data being tracked and evaluated are accurate and germane.

5.5.4 Auditing as an Environmental Risk Management Tool

Audits help minimize risk by assessing, monitoring, and controlling potential legal and financial liabilities. Legal liabilities stem primarily from the potential sanctions associated with the enforcement of regulatory violations. Most regulatory programs currently in place all have fines, penalties, and other sanctions available for timely and expeditious regulatory enforcement. Under the Clean Air Act Amendments of 1990, for example, sanctions may range from on-the-spot fines of up to $5,000 for air permit violations to $25,000-a-day penalties for more egregious violations. Even to a large company, such fines and penalties start to have an impact very quickly.

In addition to the monetary liability associated with regulatory non-compliance, there is also the specter of criminal liability. Under the new United States Department of Justice sentencing guidelines, knowing and willful violation of certain regulations can result in criminal prosecution, and possibly imprisonment. There are few things that focus the mind of a corporate executive as clearly and quickly as the imminent threat of imprisonment.

Criminal prosecution notwithstanding, legal liabilities associated with environmental, health, and safety compliance are driven largely by the threat of costly toxic tort litigation. If a spill or release of a regulated material impacts someone else's property or causes injury or illness, toxic tort litigation will likely result.

Financial liabilities represent the cost associated with bringing a facility into compliance, including permitting costs, registration costs, soil and groundwater investigation costs, and, ultimately, environmental cleanup costs. They also include the cost associated with hiring internal staff, consultants, lawyers, and other experts to address these compliance issues, and the opportunity costs associated with the disruption of operations associated with addressing environmental risk management. Finally, there is a real but hard-to-quantify financial risk associated with a negative environmental image. Environmental reputation can impact a company's standing with customers and investors, and can be much more difficult and costly to regain than a leaking tank or an emission exceedence.

5.5.5 Conducting the Audit

Prior to undergoing the compliance audit, the company should always meet with the consultant to review business activities at the site and to explain the company's environmental management program and procedures. The company should also go over the process that the consultant will utilize in conducting the audit. In general terms, the compliance audit would include the following basic steps:

1. *Pre-audit* - Select the consultant, schedule the audit, and gather background information for the consultant.

2. *Site Investigation; Initial Findings* - The consultant meets with company managers to go over the basic routine of the audit. The consultant gathers written records from the company files; conducts interviews with company employees; makes a visual review of the company site; reviews procedures for the handling, storage and disposal of wastes; reviews emergency planning and preparedness procedures; reviews recordkeeping and reporting procedures; takes air, water and soil samples; notes findings on an audit checklist; and then reviews the initial findings with company managers.

3. *Audit Report; Action Plan* - The consultant reviews the information gathered from the site investigation. The consultant then recommends an action plan for bringing the facility into regulatory compliance by specified dates. The company should designate responsible employees to carry out the action plan and periodically review the plan to make sure that each component is fully implemented.

The audit is designed to alert the company to areas of noncompliance so that expensive fines and penalties may be avoided. The company may lack certain required permits. Improper labeling may be present. Inadequate training of employees may be discovered. Effluent or emissions levels may need to be reduced. Hazardous waste minimization may need improvement. Better recordkeeping practices may be necessary. Each problem area must be addressed and corrected in a timely manner. Violations of laws and regulations must be remedied as quickly as possible.

Even under the safest of conditions, accidents will occur; however, regulatory compliance decreases the probability of pollution problems at the site. Unlike the sudden spill or release, the company has complete control over regulatory compliance. Routine compliance audits can help to maintain compliance with regulatory standards, and should be part of every company's environmental management program.

5.5.6 Pre-Audit Activities

Preliminary activities include determining the scope of the audit, assembling the proper team, and collecting as much advance information as possible. The scope of an audit will determine how the audit will be structured and performed. It can range from an environmental audit of a single plant to an integrated audit of a multinational corporation's worldwide facilities. Assembling the proper team is another critical component. The team members should be effective auditors and possess the right combination of training and experience needed to audit the specific subject.

Collecting information in advance is not often easy, but it helps dramatically in the long run. In many audits, this can be accomplished by submitting questionnaires or checklists to the facility in advance and reviewing the completed forms before the audit is undertaken. In addition, it is useful to collect and review as much documentation on the facility, such as historical assessments, regulatory files, and recent mandatory submissions, to enable the auditors to take maximum advantage of their time at the site.

5.5.7 Site Investigation

The site-based component of an audit will vary widely from facility to facility. Typically, though, on-site activities consist of a kickoff meeting before the work begins, a review of the documentation that is available on file at the facility and that has not been provided in advance, interviews with the appropriate facility personnel, and a thorough visual inspection of the entire facility. In addition, it is advisable to conduct daily briefings for each day that the auditors are on-site, and a close-out meeting before the auditors leave the site.

Unlike the typical transactional assessment, where the findings hinge largely on the site inspection, the compliance audit tends to depend more on sedentary work, whether it is the file review, the checklist completion, the interview, or the briefings. In this respect, an environmental audit is similar to a financial audit in that most of it takes place in an office or a conference room. Nevertheless, the site inspection will always remain a critical component of an audit, and it is the part that most requires the keenest, most experienced auditor's eye.

5.5.8 Initial Findings and Recommendations

Generally, once the audit is completed, a draft report is prepared, and findings are presented to the plant managers; the corporate environmental, health, and safety managers; the corporate counsel; or some combination of the above. Once that has been done, the auditors and the corporate team will usually develop a remedial strategy jointly to ensure that issues raised by the audit are thoroughly and completely addressed, and that the compliance issues raised during the audit no longer represent a potential liability. In many cases, the audit close-out process will include recommendations for periodic spot-checking and updating.

5.5.9 Audit Report

The audit report is the ultimate deliverable of any environmental, health, and safety audit, and the findings and recommendations are the essence of the audit. Therein lies the double-edged sword of any audit - financial or environmental. On one hand, it minimizes risk by providing a facility or company with a comprehensive documentation of its environmental compliance status. On the other hand, the audit report presents a greater risk if the findings are inaccurate or misleading, or if the recommendations are not followed. An audit report may be a discoverable document in a subsequent enforcement action or litigation, and the audit report may document reportable regulatory obligations that the facility has not met or is unable to meet.

The apparent dilemma of the audit report being both the ultimate asset and ultimate liability of the environmental, health, and safety audit, however, is not insuperable. Its resolution, though, requires an understanding of these issues, and the establishment of the ground rules during the preliminary activities. The protection of the confidentiality of the audit findings and the audit report is a complex and ongoing legal issue that may ultimately have to be resolved by the courts. Still, it is generally considered prudent to obtain legal guidance as part of the audit process.

5.6 ENVIRONMENTAL AUDITING TRENDS

In addition to understanding the basic concepts and processes of environmental auditing, it is important to recognize some of the emerging trends in the auditing field, including:

1. Self-audits;
2. Confidentiality for voluntary environmental compliance audits;
3. Certification of auditor qualifications;
4. Computerization of auditing systems; and
5. International corporate environmental audits.

5.6.1 Self-Audits

More and more organizations have begun to establish an internal program for conducting their own corporate environmental audits. This is commonly referred to as the self-audit, as opposed to the independent audit, which is performed largely by independent consultants. The advantages of self-auditing are obvious. Internal staff are usually more familiar and more experienced in the processes and operations that are being audited, and in most cases they are already on staff.

On the other hand, the corporate audit team may not be as well versed in all regulatory programs necessary for a particular audit, and corporate staff is engaged in many other activities. In those cases, it may be advantageous to hire independent consultants to perform these audits. The compromise solution that has evolved is to establish mixed audit teams consisting of corporate environmental, health, and safety personnel as well as outside experts. These mixed teams tend to maximize the level of expertise at a more modest cost.

5.6.2 Confidentiality for Voluntary Environmental
Compliance Audits

Current debate has centered on regulatory incentives for companies to institute their own self-auditing programs. Corporate self-audit privilege laws have been proposed by industry and many state government officials, while EPA has voiced opposition to confidentiality for voluntary environmental audits. Environmentalists also generally oppose such confidentiality and call for even broader corporate environmental disclosures. While EPA is reassessing its 1986 environmental auditing policy some states have enacted self-audit privilege laws which provide limited disclosure protection to information contained in voluntary corporate environmental audits. In addition, a few recent federal district court rulings have created new precedent that affords confidentiality to corporations that conduct voluntary environmental audits.[16] Finally, proposed federal environmental audit privilege legislation has prompted strong Congressional debate.

Proposed Revision of EPA's Environmental Audit Policy

EPA has been giving strong consideration to revision of its environmental audit policy, originally issued on July 9, 1986. EPA has been gathering information concerning whether additional incentives are needed at the federal level to encourage performance of environmental audits, disclosure of audit findings, and prompt correction of environmental violations uncovered during audits. Under its 1986 audit policy, EPA has encouraged the use of environmental auditing, but has emphasized that "audit reports may not shield compliance information otherwise reportable or accessible to EPA." EPA also indicated that, while the policy is not intended to preempt states from developing other approaches, there should be consistency between state and federal audit policies. Thus, EPA has been opposed to state privilege laws. EPA is asking that states delay efforts to pass legislation that would make data gathered during a voluntary environmental compliance audit off-limits to government enforcement officials.

State Audit Privilege Laws

By the end of 1994, five states had enacted environmental audit privilege laws: Colorado, Illinois, Indiana, Kentcky, and Oregon. Several other states either have proposed legislation to address the confidentiality of environmental self-audits or have addressed the issue through various policy statements. In 1995, several states had pending audit privilege legislation, including Arizona, Idaho, Iowa, Kansas, Nebraska, and Missouri.

By way of illustration, Oregon has enacted a state law that grants a limited privilege of confidentiality to companies that perform voluntary

---

16 See Reichhold Chemicals, Inc. v. Textron, Inc., 157 Fed. Rules of Decision 522 (N.D. Fla. 1994).

environmental compliance audits. Under this provision, regulated businesses and government agencies can keep the findings of voluntary "environmental audit reports" confidential, except under certain circumstances. A report may not be kept confidential if it is developed in response to requirements of state or federal law; if the privilege has been waived; if the report is created for fraudulent purposes; or if the regulated entity fails to act diligently to achieve environmental compliance.

Proposed Federal Legislation

Proposed federal legislation has further heated·up debate on the audit privilege issue. For example, during the 103rd Congress, Senator Mark Hatfield (R-Or.) introduced federal legislation in August 1994 that would give confidentiality protection to companies when conducting voluntary environmental compliance audits. Hatfield is expected to file a similar bill during the 104th Congress. The Senate bill (S. 2371), called the Environmental Audit Protection Act, was modeled after Oregon's environmental audit privilege legislation.

The proposed federal law provides that information obtained during the course of a voluntary compliance audit would remain confidential under a "limited legal privilege." The privilege would, however, become inoperable if the company waives the privilege, uses it for fraudulent purposes, or if the company fails to take prompt action to remedy any violation discovered during the audit. Further, the privilege is only applicable to information contained in the audit report; it does not extend to the violation itself. Thus, enforcement officials would not be prevented from taking action for environmental law violations.

Organizations Addressing the Audit Privilege Issue

Several different national organizations have been examining the environmental audit privilege issue, including:

1. The Coalition for Improved Environmental Audits was formed to develop a federal mechanism for encouragement of voluntary environmental audits while reducing the risk that the content of self-audit reports will be made available to environmental law enforcement officials or third parties.

2. The Compliance Management Policy Group comprised of various industry representatives is also seeking to implement an environmental audit privilege but this organization has been focusing its efforts on the enactment of state audit privilege legislation.

3. The Environmental Auditing Roundtable is another organization that is addressing the audit privilege issue. Composed of industry, regulatory, and consulting professionals, the

organization has produced an auditing Code of Ethics, an auditing legislation White Paper, and standards for conducting environmental compliance audits.

4. The Corporate Environmental Enforcement Council, made up of corporate counsel and management from a wide range of industrial sectors, has been focusing its efforts on environmental enforcement public policy issues. Organization members have been working with state business associations to promote a model state audit and disclosure proposal.

5. The ASTM Subcommittee on Environmental Regulatory Compliance Auditing has been working to provide the regulated community with standardized practices and procedures for use when assessing environmental regulatory compliance, as well as to clarify the legal issues associated with environmental compliance audits.

5.6.3   Certification of Auditor Qualifications

There has also been an effort to create a standard for auditor qualifications, thereby ensuring that only those who have been properly trained and experienced are permitted to engage in audit activities. This is largely because the environmental auditing field requires such a wide range of qualifications, including engineers, geoscientists, and regulatory affairs specialists, that it is difficult to establish universal criteria for auditor qualifications.

Although a number of private organizations and professional associations have tried to establish auditor qualification programs and certifications, to date there has been little universal consensus. Still, several states, including California and Massachusetts, do certify environmental site assessors, and various private organizations provide certification through training and correspondence courses.

Further, efforts have been made at the federal level to set up a federal program to monitor certification of professional environmental site assessors.[17] For example, in November 1993, Representative Bill Richardson (D-N.M.) introduced such a bill in the U.S. House of Representatives. The bill prposed the setting of minimum standards for organizations that train and certify professionals to perform Phase I environmental site assessments.

This House bill marked the first time that such certification was addressed at the federal level, with legislation targeting the certifying organizations rather than their individual members. The bill was designed to weed out bogus organizations that provide certification, and provide quality control in the environmental site assessor industry.

---

17 H.R. 3572, reported in 139 Cong. Rec. H10299 (Nov. 19, 1993).

The House bill proposed that EPA create an environmental certification board, which would take a careful look at how various groups run their training and certification programs, including oversight of continuing education and testing procedures, as well as codes of ethics. Certification board members would include representatives from EPA, public interest groups, the manufacturing industry, environmental consulting firms, and banking, insurance and investment sectors. Richardson introduced the bill with five co-sponsors and also approached Senator Frank Lautenberg (D-N.J.) about introducing a companion bill in the Senate that might be attached to the federal Superfund reauthorization legislation.

5.6.4 Computerization of Auditing Systems

Computerization has dramatically affected the way that environmental audits are performed. Regulatory agency databases, auditing checklists, even the regulations themselves, are now available to environmental auditors via computer. Many environmental audit teams even bring notebook or laptop computers to the site, adding an "on-line" information access component to the audit.

5.6.5 International Corporate Environmental Audits

One of the fastest-growing areas in the auditing field is the international audit. Many multinational corporations have been establishing a single standard for environmental compliance at all worldwide facilities. This enables the corporate leadership to ensure that all facilities are meeting the same stringent guidelines, and that all facilities can be measured by the same standard.

PART III ENVIRONMENTAL INSPECTIONS

Chapter 6

Handling Environmental Inspections

6.1   INTRODUCTION

In addition to maintaining proper environmental records and reporting
spills and releases of oil and hazardous substances, as discussed in the
earlier chapters, it is essential that businesses and industrial
facilities that produce, store, dispose of, and otherwise handle
hazardous substances be prepared in the event that government officials
conduct a site inspection of company operations. This chapter outlines
various measures to undertake in successfully preparing for the dreaded
environmental inspection. Chapter 7 further describes the procedures
used by government inspections and Chapter 8 provides helpful
environmental inspection checklists to aid the company in maintaining
regulatory compliance.

6.2   PREPARING FOR ENVIRONMENTAL INSPECTIONS

Government officials often provide prior notice of their intention to conduct an environmental inspection at the company site. If prior warning is given, it is wise to request, in advance, that the inspector or inspection team provide you with a copy of any inspection checklists that will be used. The checklists will help the company to determine which areas of the facility will be inspected and what records the inspector will need to review for regulatory compliance. Regardless of whether advance notice is given, the company should nevertheless have corporate policies and procedures in place to address various concerns about environmental inspections, including:

1. Who should be involved in the inspection;

2. Under what circumstances access should be denied;

3. Whether opening and closing conferences should be required; and

4. Whether the company should require that government inspectors have search warrants or subpoenas.

Each of these considerations should be thought out well in advance so that the company is not taken by surprise, can exercise some level of control over the process, and can ensure that its employees don't panic. In order to develop a good plan of action and corporate policy for handling environmental inspections, the company must know its legal rights in regard to the government's inspection authority. These legal considerations are discussed in the next section.[1]

6.3   SCOPE OF GOVERNMENT INSPECTION POWERS

The information-gathering provisions of various environmental laws empower government officials to gain access to certain premises and records. In addition, this authority generally gives the government the authority to perform sampling at the site. Civil and criminal penalties may be imposed if government access is improperly denied. The inspection authority under various federal environmental laws is summarized here. Parallel or additional inspection powers may be granted to government officials under state environmental laws. Further, it is important to note that the U.S. EPA and state environmental agencies commonly share information obtained during inspections.

---

1 See generally Chapter 7 in R. K. Goleman et al., What to Do When the Environmental Client Calls, Second Edition (American Bar Ass'n 1994).

6.3.1 Comprehensive Environmental Response, Compensation and Liability Act (CERCLA)

CERCLA Section 104(e)[2] provides EPA officials with authority to conduct inspections. The purpose of this section is to assist in (1) determining the need for response, or choosing or taking any response action, to a release of hazardous substances, and (2) enforcement if there is a reasonable basis to believe that there may be a release or threatened release of a hazardous substance.

The premises subject to inspection are places where:

1. Hazardous substances, pollutants, or contaminants are or have been generated, stored, treated, disposed of, or transported from;

2. Any such substance has been or may have been released;

3. Such release is or may be threatened; and

4. Entry is needed to determine the need for response or appropriate response or to effect a response action.

Entry must be granted to an officer, employee, or EPA representative upon reasonable notice. Access to premises is authorized for reasonable times, such as during working hours.

Upon access, the government inspector may obtain samples. Split samples must be taken upon request of the facility owner. The inspector is also entitled to access company records for review and photocopying.

CERCLA specifically provides for issuance of an administrative compliance order or injunctive relief by federal court action to compel compliance with any authorized inspection request. The statute also provides for a $25,000 civil penalty for each day of unreasonable noncompliance. Further, the government may also assert RCRA inspection authority and assess RCRA sanctions if the material of concern is a hazardous waste in addition to being a hazardous substance under CERCLA.

6.3.2 Resource Conservation and Recovery Act (RCRA)

RCRA Section 3007[3] provides EPA officials with authority to conduct inspections. The purpose of this section is development of regulations and enforcement of existing regulations.

The facility premises subject to inspection are defined as any place where hazardous wastes are or have been generated, stored, treated, disposed of, or transported from.

---

2 42 USC 9604(e).

3 42 USC 6927.

Entry must be permitted, upon request, to an EPA officer, employee, or representative. The entry must, however, be during a reasonable time and must be commenced and completed with reasonable promptness. EPA inspectors may obtain samples and split samples must be taken if requested. Inspectors have authority to access company records and make photocopies.[4]

RCRA provides for civil penalties of up to $25,000 per day[5] as well as imposition of criminal penalties in certain circumstances.[6]

RCRA Underground Storage Tank Inspections

Under RCRA Subtitle I, the underground storage tank (UST) provisions of the statute, Section 9005 grants EPA authority to develop any regulation, conduct any corrective action study, or enforce any provision under Subtitle I.[7] The regulated premises are defined as any establishment or other place where a UST is located.

Upon request, entry must be granted to an authorized EPA representative or state implementing agency at all reasonable times. However, the inspection must be commenced and completed with reasonable promptness.

Authorized inspectors may sample any regulated substances contained in any USTs on the premises and they may monitor or test any USTs, associated equipment, contents, surrounding soils, air, surface water, or groundwater. The government inspector has authority to access and photocopy company records.

Failure to comply with an order under the RCRA UST provisions may subject the violator to civil penalties of up to $25,000 per day of violation.[8]

6.3.3 Clean Water Act (CWA)

CWA Section 308[9] provides EPA officials with authority to conduct inspections. The purpose of this section includes development of effluent limitations or performance standards and government enforcement when necessary to carry out the objectives of the statute. The premises subject to inspection are defined as the effluent source location or where records are required to be maintained.[10]

---

4 42 USC 6927(a).

5 42 USC 6928(g).

6 42 USC 6928(d).

7 42 USC 6991d(a).

8 42 USC 6991e.

9 33 USC 1318.

10 33 USC 1318(a).

Upon presentation of his or her credentials, entry must be granted to an authorized EPA representative.[11] The inspector may sample regulated effluent and may access and photocopy records.

The CWA authorizes civil penalties of up to $25,000 per day and criminal penalties of $2,500 to $50,000 per day of violation and up to three year's imprisonment for the first negligent or knowing violation. More stringent penalties may be imposed for knowing endangerment and repeat offenses.[12]

6.3.4 Clean Air Act (CAA)

CAA Section 114[13] provides EPA officials with authority to conduct inspections. The purpose of this section is to (1) develop or assist in development of any implementation plan, any standard of performance, or any emission standard; (2) determine whether any person is in violation of such performance or emission standard or any requirement of such implementation plan; or (3) carry out any provision of the CAA.[14]

The premises subject to inspection access are defined as the property or location upon which an emission source is located or which is owned or operated by someone subject to CAA requirements. Entry must be granted at reasonable times to an authorized EPA official or representative upon presentation of credentials. The authorized inspector may install and maintain monitoring equipment and sample emissions. Access must be granted to review and photocopy records.

Civil penalties of up to $25,000 per day per violation are possible. Conviction of a criminal offense is punishable by a fine pursuant to Title 18 of the U.S. Code or by imprisonment of up to five years, or both.[15]

6.3.5 Other Federal Environmental Laws

Inspection authority is likewise found in other federal environmental laws. For example, facilities subject to regulation under the Federal Insecticide, Fungicide and Rodenticide Act (FIFRA),[16] Toxic Substances Control Act (TSCA),[17] and the Safe Drinking Water Act (SDWA)[18] must review the inspection authorization provided in those laws.

---

11 See United States v. Stauffer Chemical Co., 684 F.2d 1174 (6th Cir. 1982), aff'd, 464 U.S. 165 (1984).

12 33 USC 1319.

13 42 USC 7414.

14 42 USC 7414(a).

15 42 USC 7413.

16 See FIFRA 9, 7 USC 136g.

17 See TSCA 11, 15 USC 2610.

18 See SDWA 1445, 42 USC 300j-4.

6.4 SEARCH WARRANTS

A search warrant is generally required for inspections carried out by administrative agencies.[19] Even when an environmental law authorizes inspections, a search warrant is almost always required when consent to entry is denied.

This warrant requirement is a constitutional protection against warrantless government searches and seizures of private property. Pursuant to the Fourth Amendment to the U.S. Constitution, if consent to enter private property is not given, government officials must have a search warrant unless exigent circumstances exist, or there is a judicially-recognized exception to the warrant requirement. A search warrant is the traditional criminal investigative warrant issued upon a finding of "probable cause."

A probable cause finding is, however, less stringent for issuance of an administrative inspection warrant. The right of entry created under various environmental laws provides sufficient authority for agency officials to obtain administrative inspection warrants. Refusal of consent is not a prerequisite to issuance of a warrant.

In order to obtain an inspection warrant, an environmental agency must show the following:

1. It has reasonable cause to believe that a violation has occurred or is occurring; or

2. The desired inspection is a product of an inspection plan or scheme founded on neutral criteria for the enforcement of the environmental statute.[20]

Although there seems to be some authority that warrantless administrative searches are permitted where a statute already envisions resort to federal court enforcement when entry is refused,[21] the general consensus is that an inspection warrant must be obtained.[22] Further, it has been EPA's policy to seek a warrant whenever consent to entry is denied.

---

19 See Marshall v. Barlow's Inc., 436 U.S. 307 (1978); Donovan v. Trinity Indus., Inc. 824 F.2d 634 (8th Cir. 1987).

20 Marshall v. Barlow's Inc., 436 U.S. 307, 320-25 (1978); Brock v. Gretna Machine & Ironworks, Inc., 769 F.2d 1110 (5th Cir. 1985).

21 See United States v. Mississippi Power & Light Co., 638 F.2d 899, 907 (5th Cir. 1981).

22 Public Serv. Co. v. EPA, 509 F. Supp. 720 (S.D. Ind. 1981), aff'd, 682 F.2d 626 (7th Cir.), cert. denied, 459 U.S. 1127; National-Standard Co. v. Adamkus, 881 F.2d 352 (7th Cir. 1989).

6.5   DEALING WITH THE ENVIRONMENTAL INSPECTION

Ordinarily, government officials will provide prior notification that they intend to conduct an inspection of company facilities and records, although they could just show up requesting entry for inspection purposes. In either case, the company must be prepared to handle the inspection effectively. There are number of important practical considerations when it comes to dealing with environmental inspections, which are outlined in this section.

6.5.1 Greeting the Inspector

Certain company employees should be designated to greet government inspectors upon arrival on the premises. The company may choose to instruct these employees to request identification and credentials from inspectors, as well as have them sign a visitor's logbook.

6.5.2 Written Agreement About the Scope of the Inspection

The company's policy concerning environmental inspections may include a request that inspectors provide designated employees with a summary or checklist of records and facility components that the inspector wishes to review. If possible, it is best to have the inspector agree in writing to the exact scope of the inspection, including the following items:

1. Witness interview schedules;
2. Documents to be produced;
3. Sampling protocols;
4. Confidentiality issues;
5. Return of documents; and
6. Inspector's use of photographic equipment while on the premises.

6.5.3 Safety Procedures and Equipment

It is important to advise inspectors about the safety procedures and equipment necessary to enter certain areas of the facility. The company should find out whether the inspector has undergone the required OSHA training for the types of operations under review at the facility site.

6.5.4 Deciding Whether to Deny Access

If the inspector does not have an inspection warrant, the company should consider whether to deny access. In cases where the inspector arrives without advance notice, it may make sense to deny access if the company is not properly prepared to handle the inspection at that time. If prior notification has been provided, and the inspector shows up without a warrant, the company must weigh out whether it is worth giving the inspector a difficult time by denying entry since the inspector is certain to come back soon with the necessary warrant. Keep in mind that the inspector will surely be less congenial upon his return.

It is important that employees designated to deal with the inspector be aware of any operating permits that contain express authorization for the permitting agency to conduct inspections. Conditions to many environmental permits authorize inspections, so make sure that relevant company personnel are correctly advised and have properly reviewed the facility's permits for possible inspection authorizations.

From a legal standpoint, denial of access to inspectors based solely on the lack of a warrant will not result in assessment of any civil or criminal penalties as long as an emergency situation does not exist. Access may also be denied for other reasons, including:

1. The inspector lacks necessary safety equipment or has not undergone training required under OSHA or other federal laws; or

2. The inspector is seeking entry other than during the working hours of the facility.

If access is denied for one of these reasons, it is important that the inspector be told that access will be allowed upon compliance with the company's objection.

Access may not be denied for any of the following reasons:

1. The inspector's use of cameras or videorecorders;

2. Strikes or plant shutdowns; or

3. The inspector's refusal to sign a waiver restricting liability or obligations of the facility owner or operator.

6.5.5 Valid Search Warrant

If an inspector presents a valid search warrant upon arrival and access is denied, the company may be subject to criminal penalties. Therefore, it is crucial that designated company personnel ask for a copy of the warrant and read it. Determine the scope and limits of the warrant. To the extent that the warrant is limited to certain portions of the facility, the company may deny access to the remaining parts. Further, it is important to verify that (1) the warrant has been signed by magistrate or judge; and (2) the warrant authorizes entry by the agents who have appeared at the facility to conduct the inspection. The company should also request copies of any affidavits that support issuance of the warrant, although the company does not have an absolute right to receive copies of this documentation.

6.5.6 Oversee the Inspector's Activities

It is important to keep a watchful eye on the inspector. The inspector should not be given carte blanche to peruse the facility and company files. Company personnel should maintain a reasonable degree of control over the inspection process. The inspector should have an employee escort at all times so that the inspector does not go beyond the scope of the inspection as agreed-upon in advance or provided by the terms of the inspection warrant. If possible, it is a good idea to photograph or videotape the entire inspection. If the inspector takes pictures or uses a video camera, ask for copies.

6.5.7 Participate in Employee Interviews

Inspectors should not be given free access to speak with company employees. If the inspector has a warrant, he is generally limited to seizure of documentary and tangible evidence. However, with or without a warrant, inspectors frequently will attempt to speak directly with company employees. Although the company cannot prohibit employees from speaking with inspectors, it can exercise some control over discussions between inspectors and employees by requesting that company management be present during any employee interviews. Further, personnel may be advised that they are not required to speak with inspectors and that they may refuse to answer any questions asked by government inspectors.[23]

6.5.8 Sampling and Split Sample Requests

The inspection authority of most environmental laws permits the inspecting agency to perform sampling.[24] If samples are collected, it is important to ensure that representative samples are taken and collected properly. The company should always request a split sample and perform an independent sampling analysis to verify the accuracy of the government's sampling analysis. Under most environmental laws, inspectors are not required to provide split samples unless a specific request is made. When requesting a split sample, first make sure that the split sample is equal in weight and volume. The following additional information should also be obtained:

1. Written receipts that describe the samples;

2. Description of the tests to be performed on the sample; and

3. Government test results.

---

23 It is important, however, to make sure that this advice is not given in such a way that it could be construed as forbidding employees from speaking with inspectors because in certain circumstances this might subject the company to obstruction of justice charges under 18 USC 1612.

24 See 6.3.

Finally, the company should take special care to observe valid chain of custody procedures and appropriate analytical techniques when handling all split samples.

6.5.9 Inspection Conclusion

The company should request copies of all photographs and videotapes taken by the inspector, as well as receipts for samples and a document inventory. If documents are seized pursuant to a valid search warrant, the company is entitled to an inventory of the documents and the inspecting agency must file a copy of same with the court that issued the warrant. Although the inspector will probably be reluctant to do so, it doesn't hurt to ask if he or she has any preliminary findings from the inspection. Finally, as a follow-up, it may be worth it to make a written request for copies of photographs, videotapes, sample analyses, and the field report submitted by the inspector.

Chapter 7

Environmental Inspection Procedures

7.1   INTRODUCTION

The authority to conduct inspections is set forth in Section 3007(a) of RCRA, which grants authority to inspectors to enter the premises of anyone who "generates, stores, treats, transports, disposes of, or otherwise handles or has handled hazardous wastes" and to access all records pertaining to such wastes. The EPA's inspection authority is limited in the following ways:

1. Inspectors must enter premises at reasonable times;
2. Inspections must be completed as promptly as possible;

3. Receipts must be issued for all samples collected;

4. Split samples must be provided if requested; and

5. Copies of any sample analyses must be furnished to facility owners or operators.

This chapter outlines the procedures used by government inspectors when inspecting facilities for environmental compliance with RCRA. Discussion focuses on the following steps in the inspection process:

1. Facility entry;

2. Opening discussion;

3. Review of facility operations, waste handling procedures, and records;

4. Visual inspections;

5. Inspector's documentation of observations; and

6. Closing discussion.

Chapter 6 explained the inspection process from the business or facility's point of view, with practical considerations for effectively handling environmental inspections. In this chapter, the inspection process is discussed from the point of view of the inspector. By understanding the concerns of the inspector, companies can better prepare themselves in dealing with government inspectors and the entire process.

7.2   FACILITY ENTRY

The first stage of an inspection, facility entry, requires advance consideration and the way this phase is handled will generally set the tone for the remainder of the site visit. Inspectors will have to decide how they will act once on-site and how they will respond to any obstacles encountered.

7.2.1 Inspector's Arrival

Inspections can be announced or unannounced. Regardless, inspectors need to determine an appropriate time of entry. Inspections must be conducted at a reasonable time or during normal working hours. Inspections that cannot be completed before the normal close of business will continue on the next business day, unless management does not object to completing the inspection after closing time (if the time needed to complete the inspection is short). If a facility is open continuously, or if management leaves before operations stop, inspectors may continue an inspection using their own discretion. In any event, an inspection should be completed in a timely manner.

Upon arrival, inspectors should:

1. Locate the proper official (owner, operator, or agent) as soon as possible and determine whether this official is authorized to offer assistance.

2. Present identification to the proper officials, even if it is not requested, and keep identification in sight at all times

3. Document arrival in a logbook or field notebook, noting date, time, and the names and titles of facility personnel encountered.

Generally, proper identification consists of an inspector's EPA or state agency identification card and any additional identification required by EPA Regional or state policy. Inspectors should familiarize themselves with applicable regional or state policy on identification requirements.

Inspectors may be asked on arrival to sign a log, passbook, waiver, or other form prior to entering the facility. In general, inspectors may sign logs or passbooks; they are used by facilities to keep a record of visitors to the facility and are useful in the event of a fire or other emergency.

Inspectors should not sign waivers or other legal documents if they limit the facility's liability in the event of an accident. Additionally, inspectors should not sign other legal documents limiting their rights or the owner's responsibilities while the facility visit is occurring.

7.2.2 Consent

The owner or agent in charge of a facility at the time of an inspection must give consent to the inspector to inspect the premises.

Inspectors should note that a consent to inspect may be withdrawn at any time. However, any segment of an inspection that is completed before such withdrawal remains valid. Withdrawal of consent is equivalent to a refused entry. In such an event, inspectors must secure a warrant to complete the inspection.[1]

Inspectors may observe and report on things in plain view (i.e., anything that a member of the public could be in a position to observe) even without consent to site entry. This includes observations made while on private property in areas not closed to the public (e.g., matters observed while the inspector presents identification).

---

1 Refusal of entry and use of a warrant to obtain entry are discussed in Sections 7.2.4 and 7.2.5.

During an inspection, an owner/operator may try to limit an inspector's access to portions of the facility. Limiting access to portions of the facility is similar to denying access to the facility. The appropriate response to being denied access is discussed below.

7.2.3 Denial of Access

Inspectors may be denied access for several reasons, some of which may be valid. Inspectors can reasonably be denied access if they do not have the safety equipment required by a facility (per OSHA or NIOSH requirements). In such a case, it will generally be possible to obtain access by satisfying the owner/operator's objection (e.g., by returning on another day with the required safety equipment). Inspectors do not usually need a warrant to obtain access in such cases.

Legally indefensible actions resulting in denial of access include:

1. An owner/operator refusing to allow an inspector to bring in necessary equipment (e.g., camera);

2. An owner/operator refusing an inspector access to documents;

3. An owner/operator refusing entry due to a strike and/or plant shutdown;

4. An owner/operator refusing entry due to an inspector's refusal to sign a waiver or other legal document restricting the owner/operator's liabilities or obligations.

If access is denied, the inspector will carry out the following procedures in response to the denial of access:

1. The inspector will ask the reason for denial.

2. If the problem is beyond the inspector's authority, he may suggest that the company official contact an attorney to obtain legal advice on his/her responsibility under 3007 of RCRA.

3. The inspector must not, under any circumstances, discuss potential penalties or do anything that might be construed as threatening.

4. If access is still denied, the inspector will fill out a "Denial of Access Report" (the format is set forth on the next page) and request the signature of the facility representative.

5. The inspector will then leave the premises and document any observations made pertaining to the denial, particularly any suspicion of violations.

Format for Denial-of-Access Report

DENIAL-OF-ACCESS REPORT

On _____ at _____ I was denied access into

_____ at _____
                                                                         Location

by _____
           Facility Representative's Name and Title

for the following reason(s):

<u>List here</u>:

1. _____

2. _____

_____
                                     Signed/Inspector

_____
                             Signed/Facility Representative

The facility representative, _____ , has refused

to sign this Denial-Of-Access Report. (_____ , _____ .)

6. The inspector will report all aspects of denial of access to the appropriate EPA Regional or state enforcement division to determine the appropriate action to take and to get help in obtaining a search warrant

7. A federal or state enforcement division attorney will usually assist the inspector in preparing the documentation necessary to obtain a search warrant and in arranging for a meeting with the inspector and a U.S. or State Attorney (the inspector or attorney will bring a copy of the appropriate draft warrant and affidavits to the meeting).

8. For federal inspections, the enforcement division attorney will inform the appropriate EPA Headquarters enforcement attorney or equivalent of any entry refusals and will forward copies of all papers filed.

9. The attorney will then secure a warrant and forward it to the inspector and/or the U.S. Marshal or equivalent state law enforcement authority.

7.2.4 Inspector's Use of Warrant to Gain Access

Inspectors are trained to keep the following points in mind when seeking access to a facility under a search warrant:

1. A U.S. Marshal or local law enforcement officer should accompany inspectors if the probability is high that entry will still be refused, or if the owner/operator has made threats of violence.

2. Inspectors should never attempt to make forceful entry into a facility.

3. If an owner/operator refuses entry to an inspector with a warrant and the inspector is not accompanied by a U.S. Marshal or local law enforcement officers, the inspector should leave the facility and inform an enforcement division attorney.

7.2.5 Inspection Conducted Under Warrant

The procedures for conducting an inspection under a search warrant differ from those for conducting an inspection under normal circumstances. When the inspection is conducted pursuant to a search warrant:

1. The inspection must be conducted in strict accordance with the warrant. If the warrant restricts the inspection to certain areas

or to certain records, inspectors must comply with these restrictions.

2. If sampling is authorized, all procedures must be carefully followed, including presentation of receipts for all samples taken. The facility should also be informed of its right to retain a portion of the samples obtained by inspectors.

3. If records or property are authorized to be taken, inspectors must provide receipts and maintain an inventory of all items removed from the premises.

7.2.6 Threats Made to Inspectors

The receptiveness of facility officials to an inspection will vary from facility to facility. In general, most inspections proceed without difficulty. However, in some cases, facility representatives may threaten inspectors trying to obtain entry to the facility or during the course of an inspection (e.g., when trying to obtain access to a particular portion of a facility or after an inspector suggests the existence of a violation).

Inspectors should determine the appropriate course of action for managing a threat based upon the nature of the threat and the actions of facility officials. If threatened with violence, inspectors should terminate an inspection and follow procedures presented in the section entitled "Denial of Access." In such cases, inspectors should not return to the facility unless accompanied by a U.S. Marshal or local law enforcement officer. Inspectors will probably need to obtain a warrant in these cases.

If inspectors receive threats that do not involve a threat of physical harm (e.g., a threat to call the inspector's supervisor), they will not generally need to terminate the inspection, unless the owner/operator withdraws consent or denies access in addition to making a threat. They should also be certain to note the threats in their field log.

Inspectors must avoid making any statements to facility representatives that could be construed as threatening or inflammatory.

7.3 OPENING DISCUSSION

After inspectors locate proper facility authorities and present their identification, they may find it appropriate to discuss their inspection plans with facility officials. The following sample agenda summarizes the key points that inspectors will likely bring up during an opening discussion:

* *Outline of inspection objectives*--a presentation of inspection objectives informs facility officials of the purpose and scope of the inspection and helps to avoid misunderstandings.

* *Provide management with information on RCRA*--during an initial inspection, inspectors may wish to discuss the provisions of RCRA and any new requirements that may affect the facility, as well as furnishing a copy of the Act. Acting in this manner, inspectors are regarded as sources of regulatory information and can help strengthen Agency-industry relations.

* *Order of the inspection*--discussing the sequence of the inspection will eliminate wasted time by allowing company officials time to make records available and to start intermittent operations.

* *Meeting schedules*--scheduling meetings with key company personnel saves time spent on waiting for people to become available. Inspectors should obtain business cards from all persons interviewed during the inspection.

* *Accompaniment by facility personnel*--during compliance inspections, it is helpful if a facility representative accompanies inspectors to explain operations and to answer questions.

* *Schedule a closing conference*--a wrap-up meeting should be scheduled with appropriate officials to provide a final opportunity to gather information, to answer questions, and to complete administrative duties.

* *Advise management of the availability of duplicate samples*--the facility has a right to request, and receive immediately, duplicates of any samples collected during the inspection for laboratory analysis, as well as copies of subsequent analysis results (if an enforcement case is not pending or being pursued).

* *Gather general information*--inspectors should obtain any necessary general information, such as the name and address of the chief executive officer of the facility.

* *Confidential business information*--an owner/operator should inform inspectors if and when information is confidential. If an owner/operator makes a claim of confidentiality, inspectors should provide the appropriate forms.

Holding an opening discussion immediately after receiving access to a facility may not be appropriate in all cases. Depending upon the objective of an inspection, inspectors may want to see particular operations or locations in a facility prior to an opening discussion.

For example, in an unannounced inspection of a facility with a suspected violation, an inspector may want to go directly to the site of the suspected violation to observe the violation before the owner/operator can stop, conceal, or otherwise obscure the noncomplying operation or condition.

Throughout an inspection, inspectors consider themselves investigative reporters who are searching for information that shows noncompliance with regulations. If inspectors diligently question facility personnel and observe operations, they will be able to discern inconsistencies in what they see, hear, and have previously reviewed, leading to possible findings of violations.

Inspectors will pursue inconsistencies until they are resolved. For example, if a facility is using a commercial solvent that generates a listed waste, but does not report that it is generating that waste, inspectors should determine what happens to the solvent. Questions to ask include "Where is the solvent used in the plant?" "Is it all consumed during use?" Inspectors must then decide if the facility representative's explanation is plausible, and whether it is consistent with the inspector's observations and knowledge. Inspectors will pursue inconsistencies until they are satisfied that they either constitute a violation or do not.

7.4   OPERATIONS AND WASTE HANDLING PRACTICES

Following an opening discussion, inspectors will generally ask facility representatives to describe facility operations and waste generation and management practices. Normally, inspectors will have become familiar with a facility through previous review of the facility's file. Therefore, the purposes of the discussion will be to:

   * Obtain a more detailed understanding of operations;
   * Answer any questions that inspectors may have regarding waste generation, waste flow, and waste management activities;
   * Identify any changes in operating and/or waste management practices; and
   * Identify and reconcile any discrepancies between the operations described by the facility representative and those described in the facility file.

During this discussion, inspectors usually prepare waste information sheets on each waste managed at the facility.

7.5   RECORD REVIEW

After discussing facility operations and waste handling practices, inspectors usually proceed to the record review. The record review provides inspectors with the opportunity to become thoroughly familiar with a facility and to formulate specific questions to be investigated during the visual inspection of the facility. However, the record review does not have to occur before the visual inspection. In some cases, inspection objectives may be served better if the visual inspection occurs before the record review. The visual inspection may be performed first for other reasons (e.g., availability of facility personnel or weather conditions).

RCRA inspectors are responsible for reviewing all recordkeeping that is required of the facility owner/operator. Although no standard format is required, inspectors should check for:

1. The presence of required records or plans;
2. Dates of the documents to ensure the documents are kept up to date and/or maintained for the required period; and
3. Any suspected falsification of data.

The regulatory requirements under RCRA Parts 262, 263, 265, 266, 268, 270, and 279 mandate that the following records be maintained by regulated parties.

TABLE 7A - RECORDS TO BE MAINTAINED BY REGULATED PARTIES

1. Generators:

40 C.F.R. 262.34	Job titles and personnel records, agreements with local authorities, and contingency plan.
40 C.F.R. 262.40	Manifests, biennial reports, exception reports, and waste analyses and test results (or other bases for determining the hazardous nature of a waste and its classification).
40 C.F.R. 268.7	Land disposal notification and certification.

2. Transporters:

40 C.F.R. 263.22	Manifests, shipping papers for bulk shipments by rail or water, and manifests for foreign shipments
40 C.F.R. 279.46	Tracking records for shipments of used oil.

3. Treatment, Storage, and Disposal Facilities:

General facility standards, including the following:

40 C.F.R. 265.13	Waste analysis plan
40 C.F.R. 265.15	Inspection schedule
40 C.F.R. 265.16	Job titles and personnel records
40 C.F.R. 265.51,.53	Contingency plan
40 C.F.R. 265.71-.77	Manifest system (records of manifests)
40 C.F.R. 265.73	Operating record
40 C.F.R. 265.93	Outline of ground-water monitoring plan

40 C.F.R.	265.94	Ground-water monitoring record
40 C.F.R.	265.112	Closure plan
40 C.F.R.	265.118	Post-closure plan
40 C.F.R.	268.7	Land disposal notification and certification
40 C.F.R.	268.19(d)	Special notification for characteristic wastes.

Facility-specific standards, including the following:

40 C.F.R.	265.193(i)	Annual assessment for tanks
40 C.F.R.	265.196(f)	Certification of major repairs
40 C.F.R.	265.197(2)	Contingent post-closure plan
40 C.F.R.	265.279	Land treatment, requirements for operating record and closure plan
40 C.F.R.	265.309	Landfills, requirements for operating record, contents and organizations of cells, and closure plan
40 C.F.R.	265.440(c)	Drip pad contingency plan
40 C.F.R.	265.441(a)	Drip pad evaluation
40 C.F.R.	265.441(b)	Drip pad upgrade plan
40 C.F.R.	265.443(a)	Drip pad assessment
40 C.F.R.	265.443(b)	Drip pad waste collection system cleaning
40 C.F.R.	266.42	Used oil analysis
40 C.F.R.	266.44	Used oil fuel analysis
40 C.F.R.	266.100(c)	Boiler and industrial furnace exemption for metals recovery units
40 C.F.R.	266.103(k)	Boiler and industrial furnace operating record
40 C.F.R.	266.108	Small quantity boiler and industrial furnace burner exemption waste quantity records
40 C.F.R.	266.111	Direct transfer equipment inspection records for boilers and industrial furnaces
40 C.F.R.	266.112	Boiler and industrial furnace waste residue data
40 C.F.R.	270.30	Permits, requirements for monitoring information (Subparts F & G).

-Required submittals to the Regional Administrator (see Table 7B).

4. Part A Permit Applicants (interim status TSDFs):

| 40 C.F.R. | 270.10 | Data used to complete permit applications. |
| 40 C.F.R. | 270.30 | Records of all monitoring information. |

5. Used Oil Processors and Re-Refiners

40 C.F.R.	279.55	Used oil analysis plan.
40 C.F.R.	279.56	Tracking records.
40 C.F.R.	279.57	Operating record.

-Required submittals to the Regional Administrator (see Table 7B).

6. Off-Specification Used Oil Burners

40 C.F.R. 279.65   Tracking records.
40 C.F.R. 279.66   Off-specification used oil certification.

-Required submittals to the Regional Administrator (see Table 7B).

7. Used Oil Fuel Marketers

40 C.F.R. 279.72   Analysis of used oil fuel.
40 C.F.R. 279.74   Tracking records.
40 C.F.R. 279.75   Off-specification used oil certification.

-Required submittals to the Regional Administrator (see Table 7B).

TABLE 7B - REQUIRED SUBMITTALS TO THE REGIONAL ADMINISTRATOR

40 C.F.R. Section 265.11
-EPA identification number.

40 C.F.R. Section 265.12
-Notice of date of arrival of hazardous waste from a foreign source.

40 C.F.R. Section 265.56
-In cases of releases, fires, or explosions, notification by emergency coordinator that an affected area is adequately cleaned before operations are resumed.

-Written report by emergency coordinator on emergency incident, within 15 days of incident.

40 C.F.R. Section 265.72
-Manifest discrepancy report within 15 days of receipt of waste.

40 C.F.R. Section 265.74
-Upon closure, copy of records of waste disposal locations and quantities.

40 C.F.R. Section 265.75
-Biennial report.

40 C.F.R. Section 265.93
-In cases of confirmation of analyses indicating significant increase (or pH decrease), a written notice that the facility may be affecting ground-water quality within 7 days of date of such confirmation.

-Within 15 days after above notification, specific plan for a ground-water quality assessment program at the facility.

-After determination of the above ground-water quality assessment, written report containing an assessment of ground-water quality and/or indicating a reinstatement of the indicator evaluation program.

40 C.F.R. Section 265.94
-Recordkeeping and reporting: groundwater monitoring information as specified.

-Annual reports of Section 265.75 contain results of ground-water quality assessment program.

40 C.F.R. Section 265.115
-Certification of closure.

40 C.F.R. Section 266.103
-Certifications of pre-compliance and compliance.

40 C.F.R. Section 270.110
-Permit application and amendments.

40 C.F.R. Section 279.51
-EPA identification number.

40 C.F.R. Section 279.57
-Biennial report.

40 C.F.R. Section 279.62
-EPA identification number.

40 C.F.R. Section 279.73
-EPA identification number.

While performing a record review in accordance with the applicable regulations, inspectors may encounter problems in accurately interpreting the regulations. Therefore, the Agency has made available a number of guidance documents and lists of background documents to aid both inspectors and the regulated community to comply with the recordkeeping requirements of Subtitle C.[2]

7.6 VISUAL INSPECTION PROCEDURES

In general, the visual inspection of a facility will proceed in accordance with an inspection plan or strategy that inspectors develop during inspection planning. This plan should outline, in the level of detail considered appropriate by inspectors, the operations they intend to inspect and the tentative order in which they will conduct the inspection. Inspectors may, however, determine that it is appropriate to

---

2 A complete catalog of background documents can be obtained from the RCRA-Superfund Industrial Assistance Hotline at 1-800-424-9346 (in the Washington, D.C. area, 703-412-9810).

modify a plan based upon information obtained during the record review or other factors, such as the availability of specific personnel for interviewing or the scheduled operations of waste management units to be inspected.

Step-by-step procedures for visually inspecting a facility will vary according to the facility and the objectives of the inspection. Generic checklists, which may serve to guide inspectors in performing inspections and in recording results of inspections, are provided in Chapter 8.[3]

Inspectors should conduct inspections in a way that allows them to evaluate and understand the waste flow within a facility and to determine the compliance status of each segment of the facility's waste management system.

For example, in ABC Manufacturing's plant, which generates hazardous waste, stores waste for off-site disposal, and treats some waste on-site, an inspection COULD proceed using the following steps:

1. Inspect Points of Generation and Accumulation

Determine if ABC Manufacturing has identified all hazardous waste by following the manufacturing process and identifying each point at which hazardous wastes are generated and where they are stored. Determine if accumulation points meet satellite storage area requirements, if applicable. Identify waste minimization opportunities.

2. Evaluate In-Plant Waste Transport

Determine if there is potential for mislabeling, misplacing, or mishandling wastes, and if wastes are adequately tracked to enable proper identification at storage and treatment units.

3. Evaluate Storage and Treatment Units for Compliance

Determine if wastes in units correspond to those whose points of generation have been inspected and identify the source of any other wastes in the units. Determine if any hazardous wastes are generated in the unit (e.g., treatment sludge) and evaluate the management of such waste for compliance.

The progression of steps described above enables inspectors to understand the movement and control of wastes within a facility. Inspectors will then be able to identify:

---

3 Please note, however, that EPA Regional offices and state agencies may have developed their own checklists that should be used in lieu of those provided in Chapter 8.

* Hazardous wastes that may not currently be considered hazardous by the owner/operator;
* Noncomplying procedures or management practices that are part of the facility's routine operations;
* Steps in the management process during which wastes may be mishandled or misidentified, and in which there are opportunities for spills or releases;
* Unusual situations which may be encountered during an inspection that vary from the facility's stated normal operating procedures and that may indicate potential violations.

Such a progression also allows inspectors to complete a checklist and to evaluate the facility in an organized manner, helping to ensure that all aspects of hazardous waste management activities at the facility are thoroughly inspected.

Inspections may be conducted completely on foot or, at larger facilities, partially by vehicle. In any case, inspectors should note all that is happening at the facility. Although inspectors should generally follow an inspection plan to better understand waste generation and management within a facility, they should not feel compelled to adhere to their original inspection plan or route. Rather, they should feel free to diverge from their original plan to further investigate any observations that may uncover potential violations or environmental hazards.

As stated earlier, inspectors should maintain control of the pace and direction of an inspection. They should ask relevant questions of both the facility representative guiding them through the facility and of other personnel. By questioning diverse personnel, inspectors may identify inconsistencies in explanations of procedures or operations that could indicate possible noncompliance that they should further investigate, and get an indication of the adequacy of the personnel training program. Inspectors should record answers to questions and observations in a field log or notebook.[4]

Inspectors should be careful to remain oriented during the tour of a facility so that they can accurately note locations of waste management areas, possible release points, potential sampling locations, and so on. At larger facilities, inspectors should carry a map or plot plan in order to note locations and maintain their orientation.

7.6.1 Use of Inspection Checklists

As previously discussed, inspectors should complete as much of any applicable checklists as possible in the facility office, generally during the record review, prior to visually inspecting the facility (unless the objectives of the inspection or other circumstances dictate

---

4 See 7.7.

that the visual inspection occur before the record review). Inspectors should leave blank those sections of checklists that require visual inspection to complete.

During the visual inspection, inspectors should complete those sections of checklists requiring visual inspection. However, completing these sections is not the sole purpose of a visual inspection, and the inspector must not limit the visual inspection to only completing the checklist. Inspectors should be aware of, and investigate, all relevant waste generation and management activities throughout the facility, and note what is happening around them as they tour the facility. If inspectors conduct visual inspections in ways which allow them to understand how wastes are generated, transported, and managed at the facility, they should be able to complete the applicable checklists easily during the inspection and obtain other important information.

7.6.2 Determining the Need for Sampling and Identifying Sampling Points

Inspectors do not routinely conduct sampling as part of Compliance Evaluation Inspections (CEIs) at interim status and permitted facilities. Rather, they generally perform sampling during inspections in support of case development, which normally occur after potentially noncomplying conditions or criminal activities have been identified during a CEI or through some other means. Sampling procedures to be followed during case development inspections/evaluations are provided in detail in the *Technical Case Development Guidance Document*, OSWER Dir. 9938.3 (1988), available from EPA's Office of Waste Programs Enforcement.

Additional information on sampling is provided in several EPA publications, including:

Test Methods for Evaluating Solid Waste, Physical/Chemical Methods, EPA Office of Solid Waste, Pub. No. SW-846, July 1982, as amended (Update I - April 1984; Update II - April 1985); and

Characterization of Hazardous Wate Sites - A Methods Manual, Volume II: Available Sampling Methods, EPA Pub. No. 600/4-84/075 (April 1985).

If inspectors are to conduct sampling during a CEI, they will determine this, or be so informed, during inspection planning. Inspectors should refer to the above-mentioned manuals during inspection planning to obtain information on preparing sampling plans, taking samples, preserving samples, splitting samples with an owner/operator, and completing chain-of-custody requirements.

7.6.3 Conditions Indicating Need for Future Sampling

Although inspectors will not usually perform sampling during CEIs, they should be aware of, and identify, potential sampling requirements that may need to be fulfilled in future inspections, particularly in cases where an inspector has identified potentially noncomplying conditions or criminal activity. In these cases, it is possible that case development inspections/evaluations will need to be performed at the facility in the future. Some conditions indicating a possible need for future testing include:

1. The owner/operator is handling a potentially hazardous waste as a nonhazardous waste--sampling may be required to verify that the waste is hazardous or nonhazardous.

2. In-plant waste handling practices indicate that mislabeling/ misidentification of waste is likely to occur, or that wastes may vary significantly in characteristic over time and be mismanaged as a result--sampling may be required to demonstrate that the facility is mislabeling or misidentifying wastes.

3. There is visible or other observable evidence of possible releases of hazardous wastes from waste management units, satellite storage areas, waste generating areas, and so on--sampling media and wastes may be required to demonstrate that a release has occurred or is occurring.

4. Wastes may be managed improperly (i.e., in an inappropriate treatment or disposal unit)--sampling may be required to verify that the correct wastes are being managed in the facility's various waste management units.

To facilitate any future sampling, inspectors may identify the media or wastes to be sampled, the physical locations at which sampling should occur (e.g., the location of a possible release), the steps within a treatment process to sample, the physical characteristics of the medium to be sampled (e.g., sludge, granular solid), and other relevant information.

7.6.4 Observations for Follow-Up Case Development

In all cases, inspectors should accurately and validly document all observations that may lead to or support further case development activities. They should record in their notebooks any and all observations made during an inspection and, where appropriate, use other forms of documentation (e.g., photographs) to further record potentially noncomplying conditions. Observations of potentially noncomplying conditions or criminal activity made by inspectors during CEIs may result in the initiation of enforcement actions.

7.7 DOCUMENTATION

Documentation refers to all printed and mechanical media produced which inspectors copy or take to provide evidence of suspected violations. Information that inspectors collect during an inspection should only be recorded by the following means: field notebooks, checklists, photographs, maps, and drawings. Inspectors are discouraged from recording information on other loose papers because loose papers may be easily misplaced and the information on them discredited during enforcement hearings. Since the government's case in a formal hearing or criminal prosecution often hinges on evidence that inspectors gather, it is crucial that proper documentation and document control are observed. Accordingly, inspectors must keep detailed records of inspections, investigations, photocopies, photographs taken, and so on, and thoroughly review all notes before leaving a site.

7.7.1 Document Control

Document control ensures accountability for all documents when an inspection is completed. Accountable documents include items such as logbooks, field data records, correspondence, sample tags, graphs, chain-of-custody records, bench cards, analytical records, and photos. To ensure proper document control, inspectors should label each document with a serialized number. The document should be listed, with the number, in a project document inventory which is assembled upon completion of an inspection. Waterproof ink should be used to record all data on serialized, accountable documents.

7.7.2 Field Notebook

In keeping field notes, inspectors should maintain a legible daily diary or field notebook containing accurate and inclusive documentation of all inspection activity, conversations, and observations. This field notebook should also include any comments, as well as a record of actual or potential future sampling points, photograph points, and areas of potential violation. The diary or field notebook should contain only facts and observations because it will form the basis for later written reports and may be used as documentary evidence in civil or criminal hearings. Notebooks used for recording field notes should be bound and have consecutively numbered pages. A separate notebook should be used for each facility inspected, in case the notebook has to be made available to the owner/operator and/or his or her attorney as part of a legal action (e.g., through discovery). Because field notebooks may be made available to owner/operators and their attorneys, inspectors should be careful to avoid recording potentially embarrassing notes or notes which may weaken any future enforcement action.

7.7.3 Checklists

In general, inspectors should use checklists in conjunction with field notebooks to record inspection observations. However, EPA Regions or states may have different policies on the use of checklists, and inspectors should follow their applicable EPA Regional or state policy.[5] Also, some inspectors may not be comfortable with checklists and should find a mechanism for recording information consistent with his or her style.

As discussed in Section 7.6, inspectors should remember that checklists are only a tool for organizing, conducting, and recording the results of an inspection; they should not limit the scope of an inspection in any way. Furthermore, inspectors should not rely on checklists as a substitute for knowledge and understanding of the regulations. Inspectors should be observant of the general operation of a facility, waste management practices, and potentially regulated activities not covered by checklists (e.g., new activities of which they were not aware in planning the inspection) as they perform the record review and visual inspection.

Inspectors should generally limit the scope of comments on a checklist to checking the relevant answers, although more extensive comments may be made if no alternative record is available for noting observations. Comments or observations on checklist answers should be recorded in the field notebook, where there is adequate room for explanations, sketches, and the like, to expand upon checklist answers.

7.7.4 Photographs

Photographs provide the most accurate documentation of inspectors' observations, and inspectors can use this significant and informative source for review prior to future inspections, at informal meetings, and at hearings. Documentation of a photograph's origin is crucial to its validity as a representation of an existing situation. Inspectors should note, in a field notebook or on a facility map, the following information about each photograph they take:

* Date
* Time
* Number of the photo on the roll
* Type of film, lens, and camera used
* Signature of photographer
* Name and ID number of site
* General direction faced by inspector when taking photograph

---

5 Chapter 8 provides checklists for use during RCRA Compliance Evaluation Inspections. In some cases, however, EPA Regions or states may have preferred checklists that should be used instead of those checklists.

* Location of checkpoint on site
* Other comments (e.g., weather conditions).

Inspectors are advised to limit their comments to these pertinent facts because any discussion of the photograph in terms of its content could jeopardize its value as evidence.

Inspectors may select the type of camera they will use, although 35mm single lens reflex cameras are most common. Inspectors should also note that photographs taken with a telephoto lens may not be admissible evidence as these lenses may distort the scale of the photo or image. When taking photos, inspectors should include a ruler, or other appropriate item, in the photograph to show the scale of a photographed object.

7.7.5 Video Cameras

If inspectors have video cameras available to them, they are well-advised to employ them as an excellent means of documentation. EPA anticipates that video cameras may gradually become standard equipment to be used on inspections. Video cameras have the unique ability to capture verbal and visual inputs simultaneously, thereby providing a more comprehensive picture of a facility. Inspectors should be sure to display the date and time of their recording in the video itself.

EPA has not developed a uniform policy pertaining to the use of video cameras on inspections. Certainly, some facility owner/operators may raise objections to their use by inspectors. Thus, inspectors need to know the specific policy of their EPA Regional office.

7.7.6 Maps and Drawings

Schematic maps, drawings, charts, and other graphic records can be useful in documenting violations. They can provide graphic clarification of a particular site relative to the overall facility; demonstrate spill or contamination parameters (e.g., the size of a contaminated area) relative to the height or size of objects; and other information that, in combination with other documentation, can produce an accurate and complete evidence package.

Maps and drawings should be simple and free of extraneous details. Basic measurements should be included to provide a scale for interpretation, and compass points should be included. Generally, maps should also be used to show where photographs were taken, and in what direction; photo locations can be shown on the map using the roll number, exposure (photo) number, and a direction arrow.

7.8   CLOSING DISCUSSION

Facility officials are usually anxious to discuss the findings of an inspection before inspectors leave. Inspectors should hold a closing meeting or conference for the presentation and discussion of preliminary inspection findings. During this meeting or conference, inspectors can answer final questions, prepare necessary receipts, provide regulatory information, and request the compilation of data that were not available at the time of the inspection. Inspectors should also be prepared to discuss general follow-up procedures, such as how results of the inspection will be used and what further communications the EPA Regional Office or state may have with the facility. Inspectors should conduct closing conferences in accordance with any applicable guidelines established by the EPA Regional Administrator or state director.

When conducting a closing discussion, inspectors should:

1. Review inspection notes and checklists in private prior to the closing discussion. Inspectors may need to take time to refer back to applicable federal or state standards, call their supervisors, talk with Regional or Headquarters counsel, or call the RCRA-Superfund Industry Assistance Hotline to obtain a clear interpretation of the regulations as they apply to the specific conditions at the facility.

   In general, at this point, inspectors should:

   a. Identify any questions that remain to be asked of facility officials. These may include questions raised during the visual inspection that need clarification and questions concerning potential violations uncovered during the inspection of which the facility representative is unaware.

   b. Determine which inspection results to discuss with the facility representatives and how to approach the discussion (i.e., how definitively to present results). Of course, all inspection findings are preliminary until reviewed by an inspector's supervisor. However, inspectors should be prepared to discuss all obvious violations of rules observed during the inspection forthrightly. Still, they should not suggest that an owner/operator of a facility has committed a criminal violation or that civil or criminal action will be taken. Inspectors may not want to discuss tentative findings when there is doubt that a violation has occurred and where they will need to conduct further review of facility conditions, regulations, and guidance to determine compliance.

c. Anticipate questions that may be asked by the facility representatives and determine how to respond. Obviously, questions that may be asked will largely depend on inspection results. Inspectors can anticipate that a facility representative may challenge specific results, ask for clarifications of rules or results, and request help in understanding how to respond to or correct conditions of noncompliance. Inspectors should be prepared to answer all questions within their ability, authority, and knowledge, and to defer answering questions that they cannot answer with certainty. Inspectors should tell the facility representative how they will follow up on deferred questions, and may refer the representative to appropriate EPA or state officials for answers to questions beyond their authority.

2. After completion of the first step, meet with the facility representatives to ask questions, review results, and answer their questions. When presenting results, inspectors should inform the facility representative that all inspection results are preliminary and that the overall compliance status of the facility will be determined after review of inspection results with supervisory personnel and the issuance of an inspection report.

In conducting a closing meeting, it is essential that inspectors maintain a professional, courteous demeanor, even though the attitude of facility representatives may not be cordial. Because inspectors are often the only contact point between EPA or a state agency and the regulated industries, they should be keenly aware of opportunities to maintain and improve agency-industry relations. The closing conference provides a good opportunity for inspectors to offer various kinds of help to facility officials, within appropriate limits. Having just completed an inspection, inspectors will have first-hand knowledge of existing problems and solutions.

Chapter 8

Environmental Inspection Checklists

8.1   GENERAL SITE INSPECTION INFORMATION FORM
8.2   GENERAL FACILITY CHECKLIST
8.3   AIR EMISSIONS CHECKLIST
8.4   CONTAINERS CHECKLIST
8.5   GENERATORS CHECKLIST
8.6   GROUNDWATER MONITORING CHECKLIST
8.7   HEALTH & SAFETY CHECKLIST
8.8   LAND DISPOSAL RESTRICTIONS CHECKLIST

When conducting inspections, most environmental agencies use their own internal checklists for evaluating regulatory compliance. The checklists will, of course, vary depending on the types of pollutants generated, environmental media exposed, facility operations evaluated, and environmental records examined by the inspectors. Some of these environmental agency inspection manuals are available to the regulated community.[1] By reviewing the procedures and checklists contained in various environmental agency inspection manuals, businesses and industrial facilities can more effectively monitor company operations to determine compliance with regulatory requirements.

In this chapter, various environmental inspection checklists have been selected and reproduced to provide practical insight into how EPA inspections are conducted and what inspectors may be evaluating for regulatory compliance. Specifically, illustrative checklists are provided that EPA inspectors commonly use when inspecting facilities for hazardous waste violations under the Resource Conservation and Recovery Act (RCRA). Although these checklists focus on regulatory compliance with RCRA, they are good general examples of the types of information inspectors are seeking when conducting inspections. These checklists should prove especially helpful to companies when performing self-audits of their hazardous waste management practices. Future updates to this book will contain additional checklists from other environmental agency inspection manuals.

---

1 See, for example, U.S. EPA, RCRA Inspection Manual (1993 edition), OSWER Directive 9938.02b, PB94-963605 (October 1993).

1. General Site Inspection Information Form

A. SITE NAME   B. STREET (or other identifier)

C. CITY   D. STATE   E. ZIP CODE   F. COUNTY NAME

G. SITE OPERATOR INFORMATION
1. Name   2. Telephone Number

3. Street   4. City   5. State   6. Zip Code

7. Facility Contact/Telephone No.   8. Responsible Official/Telephone No.

H. SITE DESCRIPTION

I. TYPE OF OWNERSHIP
___ 1. Federal   ___ 2. State   ___ 3. County   ___4. Municipal   ___5. Private

J. FUNCTION
___ 1. Generator   ___ 2. Transporter   ___ 3. Treatment   ___4. Storage   ___5. Disposal

K. REGULATORY STATUS
___ 1.   Interim Status   ___ 3. Part B Permit Application Submitted

___ 2.   Permitted Facility   ___ 4. Part B Permit Application in Preparation

L. INSPECTOR INFORMATION
1. Principal Inspector Name   3.   Organization

2. Title   4.   Telephone No. (area code and No.)

M. INSPECTION PARTICIPANTS

1.	6.
2.	7.
3.	8.
4.	9.
5.	10.

2. General Facility Checklist

Section A - General Facility Standards (40 CFR 264/5 Subpart B)	Yes	No

1. Does facility have EPA Identification No.? (§§264/5.11)   ___ ___
   a. If yes, EPA I.D. No. _____
       If no, explain _____
       _____

2. Has facility received hazardous waste from a foreign source?   ___ ___
   (§§264/5.12)
   If yes, has it filed a notice with the Regional Administrator?

Waste Analysis

3. Does facility maintain a copy of the waste analysis plan   ___ ___
   on-site? (§§264/5.13)

   a. If yes, does it include:
       1. Parameters for which each waste will be analyzed?   ___ ___
           (§§264/5.13(b)(1))
       2. Test methods used to test for these parameters?   ___ ___
           (§§264/5.13(b)(2))
       3. Sampling method used to obtain sample?   ___ ___
           (§§264/5.13(b)(3))
       4. Frequency with which the initial analyses   ___ ___
           will be reviewed or repeated?
           (§§264/5.13(b)(4))
       5. (For off-site facilities) waste analyses that
           generators have agreed to supply?
           (§§264/5.13(b)(5))   ___ ___
       6. (For off-site facilities) procedures which are used to
           inspect and analyze each movement of hazardous
           waste, including: (§§264/5.13(c))   ___ ___
           a. Procedures to be used to determine the
               identity of each movement of waste
           b. Sampling method to be used to obtain
               representative sample of the waste to be
               identified.

		Yes	No
4.	Does the facility provide adequate security through: (§§264/5.14)		
	a. 24-hour surveillance system (e.g., television monitoring or guards)?	——	——

OR

		Yes	No
b. 1.	Artificial or natural confining barrier around facility (e.g., fence or fence and cliff)? (§§264/5.14(b)) Describe:	——	——

AND

		Yes	No
2.	Means to control entry through entrances (e.g., attendant, television monitors, locked entrance, controlled roadway access)? (§§264/5.14(b)(2)(ii)) Describe:	——	——

General Inspection Requirements

5. Does the owner/operator maintain a written schedule at the facility for inspecting: (§§264/5.15)

		Yes	No
a.	Monitoring equipment?	——	——
b.	Safety and emergency equipment? (§§264/5.15(b))	——	——
c.	Security devices:	——	——
d.	Operating and structural equipment? .	——	——
e.	Types of problems with equipment:		
1.	Malfunction (§§264/5.15(a))	——	——
2.	Operator error	——	——
3.	Discharges	——	——

		Yes	No
6.	Does the owner/operator maintain an inspection log? (§§264/5.15(d))	——	——
a.	If yes, does it include:		
1.	Date and time of inspection?	——	——
2.	Name of inspector?	——	——
3.	Notation of observations?	——	——
4.	Date and nature of repairs or remedial action?	——	——

b.   Are there any malfunctions or other deficiencies not corrected? (Use narrative explanation sheet.) (§§264/5.15(c))

Personnel Training   Yes   No

7.   Does the owner/operator maintain personnel training records at the facility? (§§264/5.16)
How long are they kept?_____

        a.   If yes, do they include:

                1.   Job title and written job description of each position? (§§264/5.16(d))
                2.   Description of type and amount of training?
                3.   Records of training given to facility personnel?

Requirements for Ignitable, Reactive, or Incompatible Waste

8.   Does facility handle ignitable or reactive wastes? (§§264/5.17)

        a.   If yes, is waste separated and confined from sources of ignition or reaction (open flames, smoking, cutting and welding, hot surfaces, frictional heat), sparks (static, electrical, or mechanical), spontaneous ignition (e.g., from heat-producing chemical reactions), and radiant heat?

                1.   If yes, use narrative explanation sheet to describe separation and confinement procedures.
                2.   If no, use narrative explanation sheet to describe sources of ignition or reaction.

        b.   Are smoking and open flame confined to specifically designated locations?
        c.   Are "No Smoking" signs posted in hazardous areas?
        d.   Are precautions documented (Part 264 only)? (§264.17(c))

9.   Are containers leaking or corroding? (§§264/5.171)

10.   Is there evidence of heat generation from incompatible wastes?

Section B - Preparedness and Prevention (40 CFR 264/5 Subpart C)

1.   Is there evidence of fire, explosion, or contamination of the environment?

    If yes, use narrative explanation sheet to explain.

2.   Is the facility equipped with: (§§264/5.32)

        a.   Internal communication or alarm system?

                (i)   Is it easily accessible in case of emergency? (§§264/5.34)

		Yes	No
b.	Telephone or two-way radio to call emergency response personnel? (§§264/5.32(b))	——	——
c.	Portable fire extinguishers, fire control equipment, spill control equipment, and decontamination equipment? (§§264/5.32(c))	——	——
d.	Water of adequate volume for hoses, sprinklers, or water spray system? (§§264/5.32(d)) Describe source of water:	——	——

3. Is there sufficient aisle space to allow unobstructed movement of personnel and equipment? (§§264/5.35) —— ——

4. Has the owner/operator made arrangements with the local authorities to familiarize them with characteristics of the facility? (Layout of facility, properties of hazardous waste handled and associated hazards, places where facility personnel would normally be working, entrances to roads inside facility, possible evacuation routes.) (§§264/5.37) —— ——

5. In the case that more than one police or fire department might respond, is there a designated primary authority? (§§264/5.37(a)(2)) —— ——
   a.   If yes, name primary authority:

6. Does the owner/operator have phone numbers of and agreements with State emergency response teams, emergency response contractors, and equipment suppliers? (§§264/5.37(a)(3)) —— ——

   a.   Are they readily available to all personnel? —— ——

7. Has the owner/operator arranged to familiarize local hospitals with the properties of hazardous waste handled and types of injuries that could result from fires, explosions, or releases at the facility? (§§264/5.37(a)(4)) —— ——

8. If State or local authorities decline to enter into the arrangements called for under §§264/5.37, is this entered in the operating record? (§§264/5.37(b)) —— ——

Section C - Contingency Plan and Emergency Procedures (40 CFR 264/5 Subpart D)

1. Is a contingency plan maintained at the facility? (§§264/5.51) —— ——

   a.   If yes, is it a revised SPCC Plan? (§§264/5.52(b)) —— ——
   b.   Does contingency plan include:

       1.   Arrangements with local emergency response organizations? (§§264/5.52(c)) —— ——
       2.   Emergency coordinator's names, phone numbers, and addresses? (§§264/5.52(d)) —— ——
       3.   List of all emergency equipment at facility and descriptions of equipment? (§§264/5.52(e)) —— ——
       4.   Evacuation plan for facility personnel? (§§264/5.52(f)) —— ——

2. Is there an emergency coordinator on site or on call at all times? (§§264/5.55) Yes ____ No ____

Section D - Manifest System, Recordkeeping, and Reporting (40 CFR 264/5 Subpart E)

1. Does facility receive waste from off-site? (§§264/5.71(a)) ____ ____

    a. If yes, does the owner/operator retain copies of all manifests? ____ ____

        1. Are the manifests signed and dated and returned to the generator? ____ ____
        2. Is a signed copy given to the transporter? ____ ____

2. Does the facility receive any waste from a rail or water (bulk shipment) transporter? (§§264/5.71(b)) ____ ____

    a. If yes, is it accompanied by a shipping paper? ____ ____

        1. Does the owner/operator sign and date the shipping paper and return a copy to the generator? ____ ____
        2. Is a signed copy given to the transporter? ____ ____

3. Has the owner/operator received any shipments of waste that were inconsistent with the manifest (manifest discrepancies)? (§§264/5.72) ____ ____

    a. If yes, has he attempted to reconcile the discrepancy with the generator and transporter? ____ ____

        1. If no, has Regional Administrator been notified? ____ ____

4. Does the owner/operator keep a written operating record at the facility? (§§264/5.73(a)) ____ ____

    a.   If yes, does it include: (§§264/5.73(b))   Yes   No

        1.   Description and quantity of each hazardous waste received?   —— ——

        2.   Methods and dates of treatment, storage, and disposal?   —— ——

        3.   Location and quantity of each hazardous waste at each location?   —— ——

        4.   Cross-references to manifests/shipping papers?   —— ——

        5.   Records and results of waste analyses?   —— ——

        6.   Report of incidents involving implementation of the contingency plan?   —— ——

        7.   Records and results of required inspections?   —— ——

        8.   Monitoring or testing analytical data? (Part 264)   —— ——

        9.   Closure cost estimates and, for disposal facilities, post-closure cost estimates? (Part 264)   —— ——

      10.   Notices of generators as specified? (§264.12(b))   —— ——

      11.   Certification of permittee waste minimization program? (§264.73(b)(9))   —— ——

      12.   Land disposal restriction records required by §268.5, §268.6, §268.7(a), and §268.8, as applicable? (§264.73(b)(10)-(16))   —— ——

5.   Does the facility submit a biennial report by March 1 every even-numbered year? (§§264/5.75)   —— ——

    a.   If yes, do reports contain the following information:

        1.   EPA I.D. number? (§§264/5.75(a))   —— ——

        2.   Date and year covered by report? (§§264/5.75(b))   —— ——

        3.   Description/quantity of hazardous waste? (§§264/5.75(d))   —— ——

        4.   Treatment, storage, and disposal methods? (§§264/5.75(e))   —— ——

        5.   Monitoring data under §265.94(a)(2) and (b)(2)? (§265.75(f))   —— ——

        6.   Most recent closure and post-closure cost estimates? (§§264/5.75(g))   —— ——

        7.   For TSD generators, description of efforts to reduce volume/toxicity of waste generated, and actual comparisons with previous year? (§§264/5.75(h))   —— ——

        8.   Certification signed by owner/operator? (§§264/5.75(j))   —— ——

6.   Has the facility received any waste (that does not come under the small generator exclusion) not accompanied by a manifest? (§§264/5.76)   —— ——

		Yes	No
a.	If yes, has he submitted an unmanifested waste report to the Regional Administrator?	—	—
7.	Does the facility submit to the Regional Administrator reports on releases, fires, and explosions; contamination and monitoring data; and facility closure? (§§264/5.77)	—	—

3. Air Emissions Checklist

Section A - Applicability (§§264/5.1030)	Yes	No

1. Does the facility have units-permitted under Part 270 or is it permitted under Part 270?   ___   ___

    a.   What is the effective date for this facility? _____

    b.   For interim status facilities, have these requirements been incorporated into Part B application submittal?   ___   ___

2. Are there any of the following separation processes at the facility:

    a.   Distillation?   ___   ___
    b.   Fractionation?   ___   ___
    c.   Thin-film evaporation?   ___   ___
    d.   Solvent extraction?   ___   ___
    e.   Air stripping?   ___   ___
    f.   Steam stripping?   ___   ___

Section B - Waste Streams

3. Are there waste streams associated with any separation processes that contain 10 ppmw or greater organic concentration? (§§264/5.1032(a))   ___   ___

    a.   If they claim waste streams below 10 ppmw, did they use proper means to determine concentration? (§§264/5.1034(d)(1 or 2))   ___   ___

    b.   Was date of initial determination before their effective date? (§§264/5.1034(e))   ___   ___

    c.   Were other analyses performed annually or upon changes in waste streams? (§§264/5.1034(e)(2 or 3))   ___   ___

Section C - Facility Emissions Rates

4. Is the hourly process vent organic emission rate greater than or equal to 3 lb/hr? (§§264/5.1032(a))   ___   ___

    Is the yearly process vent organic emission rate greater than or equal to 3.1 tons/yr? (§§264/5.1032(a))   ___   ___

			Yes	No
a.	If performance tests were made, were they done according to §§ 264/5.1034(c)?		___	___
b.	If engineering calculations were used, were they done according to §§ 264/5.1035(b)(2)(ii)?		___	___
c.	Has the owner/operator signed a statement that test conditions portray peak capacity operating conditions? (§§264/5.1035(b)(4)(iv))		___	___
d.	Were the facility emissions rates determined by the effective date?		___	___

Section D - Facility Emission Rates After Control Devices or Change in Operations

5. a. Are the process vent organic emission rates for the facility less than or equal to 3 lb/hr and less than or equal to 3.1 tons/year or are they reduced by 95%? (§§264/5.1032(a))   ___ ___

b. If performance tests were used, were they done in accordance with §§264/5.1034(c) and was the test plan in accordance with §§264/5.1035(b)(3)?   ___ ___

c. If engineering calculations were used, were they in accordance with §§264/5.1035(b)(4)?   ___ ___

d. For facilities without the control devices installed, do they have an installation plan? ((§§264/5.1033(a)(2) and 264/5.1035(b)(1))   ___ ___

e. Will the control devices be installed by 18 months after the effective date? (§§264/5.1033)   ___ ___

Section E - Reporting (§264.1036)

6. For facilities with final permits incorporating this rule, have they sent in semi-annual reports of exceedances lasting longer than 24 hours?   ___ ___

(Use individual control device worksheets to continue inspection)

Summary Sheet for Control Devices (CD)

Vent #	Control Device	CD #	On Unit #	For Vents #
	Condenser			
	Adsorber (Regen)			
	Adsorber (Nonreg)			
	Process Heater			
	Boiler			
	Catalytic Vapor Incinerator			
	Thermal Vapor Incinerator			
	Air Assisted Flare			
	Steam Assisted Flare			
	Nonassisted Flare			

Checklist
Equipment Leak Applications
Parts 264/265 Subpart BB

Section A - Applicability (§§264/5.1050)   Yes   No

1. Is the facility permitted under Part 270 or does it have units
    permitted under Part 270?   ___   ___

    a. Facility status: interim status or permitted?

    b. What is the effective date for this facility? _____ .

2. Are any of these units exempt?   ___   ___

Section B - Waste Streams (§§264/5.1063(d))

3. Are there waste streams that contain at least 10% organics by
    weight?   ___   ___

    a. Method of determination? Knowledge, ASTM Methods
        D2267-88, E169-87, E168-88, E260-85 or Method 9060 or 8240

    b. If knowledge, is it documented?   ___   ___

    c. Date of initial determination _____

    d. Dates of other analysis? Change, batch _____ .
                                                 _____
                                                 _____

4. For each waste stream that does qualify, determine fluid type
    (gas/vapor service, light-liquid service, heavy liquid service)

    a. Method for determining light liquid service

        1. vapor pressures of constituents from standard
            texts, or

        2. ASTM D-2879-86

Section C - Facility Operating Record (§§264/5.1064(g))

5. Does the facility have a list of the equipment and identification
    numbers that are affected by this rule?   ___   ___

6.   Is there a list of the ID numbers of NDE pumps, valves, and compressors with signature of owner/operator?   Yes   No

7.   Is there a list of all affected equipment by designation?

8.   Is there a list of pressure relief devices in gas/vapor service?

9.   Dates of test for no detection emission equipment?

       Background level _____

       Maximum instrument reading _____

10.   Is there a list of ID numbers for equipment in vacuum service?

11.   List of ID numbers of "unsafe-to-monitor" and "difficult-to-monitor" valves, with explanation for each and plan for monitoring or schedule.

12.   Is there a list of valves using the skip period alternative monitoring schedule, with schedule for monitoring and % leaking determined?

13.   For dual mechanical seal pumps or compressors with barrier fluid systems with sensors, is the criteria and explanation of the criteria for determining sensor failure given?

14.   Is there an analysis of design capacity, influent/effluent for each unit subject to these requirements, and an up-to-date analysis either by testing or knowledge to determine if the equipment is covered or not?

(continued)

Identification of Equipment Covered by Rule

Equipment	Equipment ID #	Waste Stream #	Fluid
Pumps			
general			
dual mechanical			
NDE (sealless)			
closed vent/control devices			
Compressors			
general			
NDE Sealless			
CV/Control Devices			
Pressure Relief Devices			
general			
CV/Control Devices			
Sampling Connecting Systems			
general			
insitu			
Valves			
general			
leakless (NDE)			
unsafe to monitor			
difficult to monitor			
alter allowable %			
alter skip period LDRP			
Open-ended valves or lines			
Flanges and other connectors			

Date of Inspection _____
Facility _____
Inspector _____

RECORDKEEPING REQUIREMENTS (§§264/5 (b)(1) and (g))

Unit Number Listed   _____
Equipment Identification Number Listed   _____
Location at Facility   _____
Type of Equipment   _____
% by weight of TOC at equipment   _____
Fluid State at Equipment   _____
Equipment Designation   _____
If Closed-Vent/Control Device Used (264/5.1064(b)
(2-4)
    – Implementation Plan   _____
    – If testing, performance test plan   _____
    – Design Documentation or Perf. Test Results   _____
If Control Device; monitoring, operating, inspection
data (264/5.1064(e))   _____

LEAK DETECTION AND REPAIR RECORDKEEPING (§§264/5.1064 (c and d))

Monitoring Equipment Number   _____
Monitoring Operators Identification   _____
Date of Visual, Audible, Olfactory Indication of
    Leak   _____
Date of Leak Detection   _____
Date of Repair Attempt   _____
Repair Methods at each attempt   _____
Leak "Above 10,000" or Above 500 above background   _____
"Repair Delayed" if after 15 days   _____
If valve, documentation for repair delay   _____
Signature of Person approving delay   _____
Expected Date of Repair   _____
Date of Successful Repair   _____

PHYSICAL INSPECTION

Visual, Audible, or Olfactory Indication of Leak   _____
Monitoring Equipment Number   _____
Correct Calibration Method   _____
Correct Monitoring Techniques Used   _____
Method 21 Results   _____
Tag on Leaking Equipment   _____
If Equipment already had tag on it:
    – Date Leak Detected   _____
    – Date of Expected Repair or Actual Repair   _____
Equipment Marked as Being in this Program   _____

---
4. Containers Checklist

Section A - Use and Management (§§264/5.171)   Yes   No

1.   Are containers in good condition?   ___   ___

Section B - Compatibility of Waste With Container (§§264/5.172)

1.   Is container made of a material that will not react with the waste
      which it stores?   ___   ___

Section C - Management of Containers (§§264/5.173)

1.   Is container always closed while holding hazardous waste?   ___   ___
2.   Is container not opened, handled, or stored in a manner which
      may rupture it or cause it to leak?   ___   ___

Section D - Inspections (§§264/5.174)

1.   Does owner/operator inspect containers at least weekly for leaks
      and deterioration?   ___   ___

Section E - Containment (§264.175)

1.   Do container storage areas have a containment system?   ___   ___

Section F - Ignitable and Reactive Waste (§§264/5.176)

1.   Are containers holding ignitable and reactive waste located at
      least 15 m (50 ft) from facility property lines?   ___   ___

Section G - Incompatible Waste (§§264/5.177)

1.   Are incompatible wastes or materials placed in the same
      containers?   ___   ___
2.   Are hazardous wastes placed in washed, clean containers when
      they previously held incompatible waste?   ___   ___
3.   Are incompatible hazardous wastes separated from each other by a
      berm, dike, wall, or other device?   ___   ___

Section H - Closure (§264.178)

1.   At closure, were all hazardous wastes and associated residues
      removed from the containment system?   ___   ___

5. Generators Checklist

		Yes	No

Section A - EPA Identification No.

1.   Does generator have EPA I.D. No.? (§262.12)   ___ ___

      a.   If yes, EPA I.D. No._____

Section B - Manifest

1.   Does generator ship waste off-site? (§262.20)   ___ ___

      a.   If no, do not fill out Sections B and D.
      b.   If yes, identify primary off-site facility(s). Use narrative
          explanation sheet.

2.   Does generator use manifest? (§262.20)   ___ ___

      a.   If no, is generator a small quantity generator (generating
          between 100 and 1000 kg/month?   ___ ___

NOTE: SQGs are only exempt if wastes are reclaimed. (See
         §262.20(e).)

      1.   If yes, does generator indicate this when sending
          waste to a TSD facility?   ___ ___

b.   If yes, does manifest include the following information?   Yes   No
(Part 262 appendix)

1.   Manifest document no.   ___   ___
2.   Generator's name, mailing address, telephone no.   ___   ___
3.   Generator EPA I.D. no.   ___   ___
4.   Transporter Name(s) and EPA I.D. no.(s)   ___   ___
5.   a.   Facility name, address, and EPA I.D. no.   ___   ___
     b.   Alternate facility name, address, and EPA
           I.D. no.   ___   ___
     c.   Instructions to return to generator if
           undeliverable   ___   ___
6.   Waste information required by DOE - shipping
     name, quantity (weight or vol.), containers (type
     and number)   ___   ___
7.   Emergency information (optional)
     (special handling instructions, telephone no.)   ___   ___
8.   Is the following certification on each manifest
     form?   ___   ___

     "This is to certify that the above named materials
     are properly classified, described, packaged,
     marked, and labeled and are in proper condition
     for transportation according to the applicable
     national and international regulations."

                                                                   ___   ___
9.   Does generator retain copies of manifests?
     (§262.40)

     If yes, complete a through e. (§262.23)

			Yes	No
a.	1.	Did generator sign and date all manifests?	—	—
	2.	Who signed for generator?		

Name_____Title_____

| b. | 1. | Did generator obtain handwritten signature and date of acceptance from initial transporter? | — | — |
| | 2. | Who signed and dated for transporter? (§262.23) | | |

Name_____Title_____

c.	Does generator retain one copy of manifest signed by generator and initial transporter? (§262.40)	—	—
d.	Do returned copies of manifest include facility owner/operator signature and date of acceptance? (§262.40)	—	—
e.	Does generator retain copies for 3 years? (§262.40)	—	—

Section C - Hazardous Waste Determination (40 CFR 262.11)

1. Does generator generate solid waste(s) listed in Subpart D (List of Hazardous Waste)?   —   —

    a.   If yes, list wastes and quantities (include EPA Hazardous Waste No.)_____

    _____

2. Does generator generate solid waste(s) listed in Subpart C that exhibit hazardous characteristics (corrosivity, ignitability, reactivity, EP toxicity)?   —   —

    a.   If yes, list wastes and quantities (include EPA Hazardous Waste No.)_____

    _____

    b.   Does generator determine characteristics by testing or by applying knowledge of processes?_____

		Yes	No

1. If determined by testing, did generator use test methods in Part 261, Subpart C (or equivalent)? ___ ___

    a. If equivalent test methods used, attach copy of equivalent methods used.

3. Are there any other solid wastes generated by generators? ___ ___

    a. If yes, did generator test all wastes to determine whether or not they were hazardous? ___ ___

        1. If no, list wastes and quantities deemed nonhazardous or processes from which nonhazardous waste was produced (use additional sheet if necessary)

_____
_____
_____

Section D - Pretransport Requirements

1. Does generator package waste in accordance with 49 CFR 173, 178, and 179 (DOT requirements)? (§262.30) ___ ___

2.   a. Are containers to be shipped leaking or corroding? ___ ___
    b. Use additional sheet to describe containers and condition.
    c. Is there evidence of heat generation from incompatible wastes in the containers? ___ ___

3. Does generator follow DOT labeling requirements in accordance with 49 CFR 172? (§262.31) ___ ___

4. Does generator mark each package in accordance with 49 CFR 172? (§262.32) ___ ___

5. Is each container of 110 gallons or less marked with the following label? (§262.32) ___ ___

Label saying: HAZARDOUS WASTE - Federal Law Prohibits Improper Disposal. If found, contact the nearest police or public safety authority or the U.S. Environmental Protection Agency.

Generator name(s) and address(es)_____
_____

Manifest document No._____

6. Does generator have placards to offer to transporters? (§262.33) ___ ___

		Yes	No

7. Accumulation time (§262.34)

   a.   Are containers used to temporarily store waste before transport?

      1.   If yes, is each container clearly dated: Also, fill out rest of No. 7 (accum. time) (§262.34(a)(2))

   b.   1.   Does generator inspect containers for leakage or corrosions? (§265.174 - Inspections)

      2.   If yes, with what frequency?

   c.   Does generator locate containers holding ignitable or reactive waste at least 15 meters (50 Feet) from the facility's property line? (§265.176 - Special Requirements for Ignitable or Reactive Wastes)

NOTE: If tanks are used, fill out checklist for tanks. (See RCRA Hazardous Waste Tank Systems Inspection Manual, OSWER Dir. No. 9938.4)

   d.   Are the containers labeled and marked in accordance with Sections D-3, -4, and -5 of this form?

NOTE: If generator accumulates waste on site, fill out checklist for General Facilities, Subparts C and D.

   e.   Does generator comply with requirements for personnel training? (Attach checklist for §265.16 - Personnel Training.)

8. Describe storage area. Use photos and narrative explanation sheet.

Section E - Recordkeeping and Records (40 CFR 262.40)

1. Does generator keep the following reports for 3 years?

   a.   Manifest or signed copies from designated facilities
   b.   Biennial reports
   c.   Exception reports
   d.   Test results

2. Where are the records kept (at facility or elsewhere)?

3. Who is in charge of keeping the records?

Name_____Title_____

6. Ground-Water Monitoring Checklist

		Yes	No
Section A - Monitoring System (40 CFR Parts 264/5 Subpart F)			

1. Does the facility have a ground-water monitoring system in operation? (§265.90) ___ ___

    a. If yes, does the system consist of: (§265.91)

        1. Minimally one upgradient monitoring well? (Part 265) ___ ___
        2. Minimally three downgradient monitoring wells? (Part 265) ___ ___

    b. Are monitoring wells cased so that the integrity of the boreholes is maintained? (§265.91) ___ ___

    c. Is a compliance monitoring system installed whenever hazardous waste constituents are detected at the compliance point ? (§264.92) ___ ___

    d. Is a corrective-action program initiated whenever the ground-water protection standard is exceeded? (§264.100(c)) ___ ___

    e. Is a detection monitoring program instituted in all other cases? (§264.98) ___ ___

2. Does the facility have a monitoring and response program? (Part 264) ___ ___

    a. If yes, is a compliance monitoring system instituted whenever hazardous constituents are detected at the compliance point? (§264.99) ___ ___

    b. Whenever the ground-water protection standard is exceeded, does facility institute a corrective-action program? (§264.99) ___ ___

    c. In all other cases, does the facility institute a detection monitoring program? (§264.99) ___ ___

Section B - Sampling and Analysis (40 CFR 265.92)

1. Does the facility obtain and analyze samples from the ground-water monitoring system? (§265.92(a)) ___ ___

2. Has facility developed and followed a ground-water sampling and analysis plan? (§265.92(a)) ___ ___

		Yes	No

a.   If yes, does this plan include procedures and techniques for: (§265.92(a))

    1.   Sample collection?   ___   ___

    2.   Sample preservation?   ___   ___

    3.   Analytical Procedures?   ___   ___

    4.   Chain-of-custody control? ___   ___

b.   Does the facility determine the concentration or value of the following parameters in ground-water samples? (§265.92(b))

    1.   Parameters characterizing the suitability of the ground water as a drinking water supply, as specified in Part 265, Appendix 3?   ___   ___

    2.   Parameters establishing ground-water quality (chloride, iron, manganese, phenols, sodium, sulfate)?   ___   ___

    3.   Parameters used as indicators of ground-water contamination (pH, specific conductance, total organic carbon, total organic halogen)?   ___   ___

c.   Has the owner/operator established initial background concentrations or values of all parameters specified above at least on a quarterly basis? (§265.92)(c))   ___   ___

d.   Has owner/operator obtained at least four replicate measurements for each sample, and has he determined the initial background arithmetic mean and variance? (§265.92(c))   ___   ___

e.   After the first year, does owner/operator sample and analyze with the following frequencies: (§265.92(d))

    1.   Samples collected to establish background water quality (from above) at least annually?   ___   ___

    2.   Samples collected to indicate contamination (from above) at least semi-annually?   ___   ___

    3.   Elevation of ground-water surface at each monitoring well at each sampling event?   ___   ___

Section C - Preparation, Evaluation, and Response (40 CFR 265.93)

1.   Did owner/operator prepare an outline of a ground-water quality assessment program? (§265.93(a))   ___   ___

    a.   If yes, did program determine the following: (§265.93(a))

		Yes	No
1.	Whether hazardous waste or hazardous waste constituents have entered the ground water?	—	—
2.	Rate and extent of hazardous waste or hazardous waste constituent migration in groundwater?	—	—
3.	Concentrations of hazardous waste or hazardous waste constituents in ground water?	—	—

b. For each well, has owner/operator calculated the arithmetic mean and variance, based on four replicate measurements for each sample, and compared the results with initial background mean? (§265.93(b))   — —

c. Has owner/operator submitted information documenting any significant increase in comparisons for upgradient wells (or decrease in pH)? (§265.93(c))   — —

d. If the comparisons for downgradient wells show a significant increase (or pH decrease), has the owner/operator obtained additional ground-water samples from those downgradient wells in which a significant decrease was detected? (Samples must be split in two, and analyses must be obtained of all additional samples to determine whether the significant difference was a result of lab error.) (§265.93(c))   — —

1. If analyses (described above) were performed, and confirmed the significant increase (or pH decrease), did owner/operator notify the Regional Administrator within 7 days?   — —

2. If analyses confirmed significant increase (or pH decrease), did owner/operator submit to the Regional Administrator within 15 days after notification (discussed above) a certified ground-water quality assessment program? (§265.93(d))   — —

a. If yes, does plan include the following:

1. Number, location, and depth of wells?   — —
2. Sampling and analytical methods for those hazardous wastes and hazardous waste constituents at the facility?   — —
3. Evaluation procedures, including any use of previously gathered ground-water quality information?   — —
4. Schedule of implementation?   — —

3. Did owner/operator implement the ground-water quality-assessment program and, at a minimum, did he determine the following: (§265.93(d)(4))

			Yes	No
	a.	Rate and extent or migration of the hazardous waste constituents in the ground water?	—	—
	b.	Concentrations of the hazardous waste in the ground water?	—	—

4. Did owner/operator submit a report to the Regional Administrator containing the requests of the assessment outlined in No. 3 above within 15 days? (§265.93(d)(5))   — —

5. Did owner/operator notify the Regional Administrator of reinstatement of indicator evaluation program upon finding that no hazardous waste or hazardous waste constituents had entered the ground water? (§265.93(d)(6))   — —

6. If owner/operator determined that hazardous waste or hazardous waste constituents entered the ground water, did he either continue to make the determinations listed in No. 3 above on a quarterly basis until final closure or ground-water quality-assessment plan was implemented prior to post-closure care, or cease to make determinations required in No. 3 above if ground-water quality-assessment plan was implemented during post-closure? (§265.93(d)(7))   — —

7. If any ground-water quality-assessment program is implemented to satisfy No. 3 above prior to final closure, has owner/operator completed the program and reported to the Regional Administrator, as outlined in No. 4 above? (§265.93(e))   — —

8. If owner/operator does not monitor at least annually to satisfy No. 3 above, does owner/operator evaluate data on ground-water elevation obtained under No. 2e in Section B above to determine whether the requirements for location monitoring wells are satisfied? (§265.93(f))   — —

   a. If evaluation shows that the requirements for monitoring wells are not satisfied, has owner/operator modified the number location, or depth of the monitoring wells to bring the system into compliance?   — —

<u>Section D - Recordkeeping and Reporting (40 CFR 265.94)</u>   Yes   No

1.   Unless owner/operator is monitoring to satisfy the requirements
      of §265.93(d)(4), does owner/operator: (§265.94(a))

      a.   Keep records of the analyses required in §265.92(c) and (d),
            the associated ground-water surface elevations required
            in §265.92(e), and ground-water surface elevations
            required in 265.93(b) throughout the active life of the
            facility and throughout post-closure?   ___   ___

      b.   Report the following information to the Regional
            Administrator: (§265.94(a)(2))

            1.   Within 15 days of analysis for each quarterly
                  sampling event, does owner/operator submit
                  results of background concentrations?   ___   ___
            2.   Does owner/operator inform the Regional
                  Administrator about any parameters that exceed
                  maximum contaminant levels listed in
                  Appendix III?   ___   ___
            3.   (Annually) Does owner/operator report
                  concentrations or values of parameters listed in
                  §265.92(b)(3) for each well, including required
                  evaluations for these parameters under §265.93(b)?   ___   ___

                  a.   Does owner/operator also identify
                        differences from initial background
                        concentrations found in the upgradient
                        wells no later than March 1 following each
                        calendar year?   ___   ___

2.   Does owner/operator submit results of the ground-water surface
      elevations under §265.93(f), along with a description of the
      response, if needed? (§265.94(a)(2))   ___   ___

3.   If ground water is monitored to satisfy requirements of
      §265.93(d)(4), did owner/operator do the following:
      (§265.94(b))

      a.   Keep records of analyses and evaluations specified in the
            plan throughout active life and post-closure?   ___   ___

      b.   (Annually, until final closure) Submit to the Regional
            Administrator a report containing the results of the
            ground-water quality assessment program, including the
            calculated rate of migration of hazardous waste or
            hazardous waste constituents by March 1 following each
            calendar year?   ___   ___

Section E - General Requirements   Yes   No

1. Does facility comply with the following requirements?
    (§264.97)

    a. Are sufficient wells installed at appropriate locations and
        depths?   ___   ___
    b. Have sampling and analysis techniques been consistent?   ___   ___
    c. Have ground-water elevation data been recorded?   ___   ___
    d. Have background concentrations been determined?   ___   ___

2. If ground water is monitored to satisfy requirements of §265.93(d)
    (4), owner/operator must:

    a. Keep records of the analyses and evaluations specified in
        the plan throughout the facility's active life, and, for
        disposal facilities, throughout post-closure.   ___   ___

    b. Report the following ground-water monitoring
        information:

        1. During the first year when initial background
            concentrations are being determined, did
            owner/operator submit values within 15 days after
            completing analysis?   ___   ___
        2. If yes, did owner/operator also submit an
            identification of any parameters whose
            concentrations exceed maximum levels in
            Appendix III?   ___   ___
        3. (Annually) Did owner/operator report
            concentrations or values of the parameters listed in
            §265.93(b)(2) for each well, along with required
            evaluations for these parameters under §265.93(b)?   ___   ___
        4. Did owner/operator also separately identify any
            significant differences from initial background
            concentrations for upgradient wells?   ___   ___
        5. Did owner/operator report on the results of ground-
            water surface elevations (and a description of the
            results if necessary) by March 1 of the following
            year?   ___   ___

Section F - Detection Monitoring Program (40 CFR 264.98)

1. Has owner/operator established detection monitoring system to
    provide reliable indications for detection releases?   ___   ___

    a. If yes, are the following components included in the
        system:

		Yes	No
1.	Background values?	___	___
2.	Determination of ground-water flow rate?	___	___
3.	Determination of ground-water compliance point semiannually?	___	___
4.	Determination of statistically significant increases over background concentrations?	___	___
5.	Notification to the Regional Administrator if there was a statistically significant increase?	___	___

Section G - Compliance Monitoring Program (40 CFR 264.99)

1. Does facility operate a compliance monitoring program?   ___   ___

   a. Does facility determine concentrations of hazardous constituents at least quarterly?   ___   ___

   b. Does facility determine ground-water flow rate and direction in uppermost aquifer annually? (§264.99(e))   ___   ___

   c. Does facility analyze samples for Appendix IX constituents annually? (§264.99(g))   ___   ___

   d. Does facility make statistically significant increases over background values? (§264.99(h))   ___   ___

   e. If there is an increase, does facility notify the Regional Administrator and establish a corrective-action program? (§264.99(h))   ___   ___

Section H - Corrective -Action Program (40 CFR 264.100)

1. Does facility follow a corrective-action program that meets the facility's permit requirements?   ___   ___

* See RCRA Ground-Water Monitoring Systems (OSWER Directive Nos. 9950.2, 9950.3, 9950.4)

7. Health & Safety Checklist

A. FACILITY NAME   EPA ID NO.

B. FACILITY ADDRESS

C. FACILITY OPERATOR INFORMATION

    1. Contact Name   2. Phone No.
    3. Address (if different from facility)

D. PROCESS UNIT DESCRIPTION (describe type and number of units)

E. TYPE OF OWNERSHIP

    __1. Federal   __2. State   __3. County   __4. Municipal   __5. Private

F. TYPE OF FACILITY

    __1. Treatment   __2. Storage   __3. Disposal

G. REGULATORY STATUS

    __1. Interim Status   __3. Part B Permit Appeal Pending (note areas
                                     of appeal)
    __2. Permitted Facility

H. PRINCIPAL INSPECTOR

    1. Name   3. Organization
    2. Title   4. Telephone No.

I. INSPECTION PARTICIPANTS
        Name   Organization   Phone No.
    1.
    2.
    3.
    4.
    5.

NOTE: All inspection participants must have current training certification in
accordance with 29 CFR 1910.120.

Regulatory Citation/Description	RCRA Checklist Item	Yes	No	Potential Violation
§§264/265.16(a)(1) Outline of Personnel Training Program	Is there an outline of the introductory training program to prepare personnel to operate or maintain the facility in a safe manner? (Attach copy of outline or summarize below.)	___	___	_____
	Is there an outline of the review training program to prepare personnel to operate or maintain the facility in a safe manner? (Attach copy of outline or summarize below.)	___	___	_____
	Does the outline describe how the training will be designed to meet actual job tasks?	___	___	_____
	Is on-the-job training used to meet this requirement?	___	___	_____
§§264/265.16(d)(1) §§264/265.16(d)(2) Job Title/Job Description	Is a job title provided for each employee whose position at the facility is related to hazardous waste management?	___	___	_____
	Is a job description provided for each employee whose position at the facility is related to hazardous waste management?	___	___	_____
§§264/265.16(c) and (d)(3) Training Content, Frequency, and Techniques	Is the personnel training program strictly classroom instruction?	___	___	_____

Regulatory Citation/Description	RCRA Checklist Item	Yes	No	Potential Violation
	Is the personnel training program strictly on-the-job training?	___	___	_____
	Does the training program combine classroom instruction and on-the-job training?	___	___	_____
	Is an annual refresher course required for personnel whose positions at the facility are related to hazardous waste management?	___	___	_____
§§264/265.16(a)(2) Training Director	Is the training program directed by a person trained in hazardous waste management? (Summarize qualifications below.)	___	___	_____
§§264/265.16(a)(2) Relevance of Training to Job Position	Are facility personnel instructed in hazardous waste management procedures (including contingency plan implementation) relevant to their positions?	___	___	_____
§§264/265.16(a)(3) Training and Emergency Response	Does the training program include the following emergency response procedures?	___	___	_____
	• Procedures for using, inspecting, repairing, and replacing facility emergency and monitoring equipment?	___	___	_____

Regulatory Citation/Description	RCRA Checklist Item	Yes	No	Potential Violation
	• Key parameters for automatic waste feed cut-off systems?	—	—	—
	• Procedures for utilizing communications or alarm systems?	—	—	—
	• Directions for responding to fires or explosions?	—	—	—
	• Procedures for groundwater contamination response?	—	—	—
	• Procedures for conducting shutdown operations?	—	—	—
§§264/265.16(b),(d)(4) and (3) Implementation of Training Program	Are all facility personnel trained within six months of their employment or assignment to the facility or transfer to a new position?	—	—	—
	Are facility personnel allowed to work unsupervised before their training program has been completed?	—	—	—
	Are records maintained which document that the required training has been given to and completed by facility personnel? (Summarize below.)	—	—	—
§§264/265.33 Testing and Maintenance of Equipment	Does the owner/operator test and maintain (as necessary to assure its proper operation in time of emergency) the following equipment:			
	• All communications or alarm systems?	—	—	—
	• Fire protection equipment?	—	—	—
	• Spill control equipment?	—	—	—
	• Decontamination equipment?	—	—	—

Regulatory Citation/Description	RCRA Checklist Item	Yes	No	Potential Violation
§§264/265.15 General Inspection Requirements	Does the owner/operator maintain a written schedule at the facility for the inspection of:			
	• Monitoring equipment?	___	___	___
	• Safety and emergency equipment?	___	___	___
	• Security devices?	___	___	___
	• Operating and structural equipment?	___	___	___
	Does the schedule identify the types of problems to look for?	___	___	___
	Is the frequency of inspection based on the possible deterioration of equipment and the probability of incident?	___	___	___
	Are areas subject to spills, such as loading and unloading areas, inspected daily?	___	___	___
	Does the owner/operator maintain an inspection log?	___	___	___
	If yes, does the log include:			
	• Date and time of inspection?	___	___	___
	• Name of inspector?	___	___	___
	• Notations of observations?	___	___	___
	• Date and nature of repairs or remedial actions?	___	___	___
	Have any malfunctions or other problems not been remedied? (Summarize below.)	___	___	___

Regulatory Citation/Description	RCRA Checklist Item	Yes	No	Potential Violation
§§264/265.35 Required Aisle Space	Does the facility maintain aisle space to allow unobstructed movement of personnel, fire protection equipment, spill control equipment, and decontamination equipment?	___	___	___
	If aisle space is not maintained, has the owner/operator demonstrated to the Regional Administrator that the space is not needed?	___	___	___
§§264/265.32 Equipment Requirements	Is the facility equipped with the following:			
	• An internal communications or alarm system capable of providing immediate emergency instruction (voice or signal) to facility personnel?	___	___	___
	• A device such as a telephone (immediately available) or handheld two-way radio capable of summoning emergency assistance from police, fire, or state or local emergency response teams?	___	___	___
	• Portable fire extinguishers?	___	___	___
	• Fire control equipment (including special extinguishing equipment such as foam, inert gas, or dry chemical)?	___	___	___
	• Spill control equipment?	___	___	___
	• Decontamination equipment?	___	___	___
	• Water at adequate volume and pressure to supply water hose streams, or foam producing equipment, or automatic sprinklers, or water spray systems?	___	___	___

Regulatory Citation/Description	RCRA Checklist Item	Yes	No	Potential Violation
§§264/265.17(a) and (b) Requirements for Ignitable, Reactive, or Incompatible Wastes	Does the facility handle ignitable or reactive waste?	___	___	_____
	If yes:			
	Does the owner/operator take the following precautions to prevent accidental ignition or reaction of wastes?			
	• Separate and protect ignitable or reactive wastes from sources of ignition or reaction (open flames, smoking, cutting, welding, hot surfaces, frictional heat, static electrical or mechanical sparks, spontaneous ignition, and radiant heat)?	___	___	_____
	• Does the owner/operator confine smoking and open flames to specially designated locations, while ignitable or reactive waste is being handled?	___	___	_____
	• Are "No Smoking" signs placed conspicuously wherever there is a hazard from ignitable or reactive waste?	___	___	_____
	Does the owner/operator have procedures in place to prevent accidental ignition or reaction of wastes? (Summarize below.)	___	___	_____
§§264/265.50 through §265.56 Contingency Plan	Does the owner/operator have a Contingency Plan, or a Spill Prevention Control and Counter measures (SPCC) Plan, or some other emergency plan, that is amended for hazardous waste management?	___	___	_____

Regulatory Citation/Description	RCRA Checklist Item	Yes	No	Potential Violation
	Is a copy maintained at the facility?	___	___	_____
	Has a copy been submitted to all local police and fire departments, hospitals, and State and local emergency response teams?	___	___	_____
	Does the plan describe the control procedures taken in the event of a fire, explosion, or release?	___	___	_____
	Does the plan describe how and when it will be implemented?	___	___	_____
	Does the plan describe arrangements agreed to by local police and fire departments, hospitals, contractors, and State and local emergency response teams to coordinate emergency services?	___	___	_____
	Does the plan list names, addresses, and phone numbers (office and home) of all persons qualified to act as emergency coordinators?	___	___	_____
	Is one person named as the primary coordinator?	___	___	_____
	Does the coordinator have the authority to commit the resources to carry out the emergency plan?	___	___	_____
	Does the plan physically describe and identify the location of all emergency equipment at the facility?	___	___	_____
	Does the plan include provisions to ensure that the equipment is cleaned and fit for its intended use before operations are resumed?	___	___	_____
	Does the plan include an evacuation plan for facility personnel?	___	___	_____

Regulatory Citation/Description	RCRA Checklist Item	Yes	No	Potential Violation
	Does the plan describe:			
	• Signal(s) to be used to begin evacuation?	___	___	___
	• Evacuation routes?	___	___	___
	Does the plan describe the methodology for immediate notification of:			
	• Facility personnel?	___	___	___
	• State or local agencies with designated response roles?	___	___	___
	Does the plan include procedures for identification of released materials?	___	___	___
	Does the plan include procedures/criteria to assess possible hazards to human health and the environment that may result from the release, fire, or explosion?	___	___	___
	Does the plan describe all reasonable measures necessary to ensure that fires, explosions, or releases do not occur, reoccur, or spread to other hazardous waste at the facility?	___	___	___
	Does the plan describe procedures to monitor for leaks, pressure buildup, gas generation, or ruptures in valves, pipes, or other equipment if the facility stops operation in response to a fire, explosion, or release?	___	___	___

Regulatory Citation/Description	RCRA Checklist Item	Yes	No	Potential Violation
§§264/265.37 Necessary Agreements with Local Authorities	Has the owner/operator made the following arrangements:			
	• Familiarized police, fire departments, and emergency response teams with the layout of the facility and associated hazards?	___	___	___
	• Designated one police and fire department with primary emergency authority when more than one might respond?	___	___	___
	• Agreements with State emergency response teams, contracts, and equipment supplies?	___	___	___
	• Familiarized local hospitals with the properties of waste handled at the facility and the types of injuries or illness that could result?	___	___	___
	• Where authorities decline to enter into such arrangements, has the owner/operator documented the refusal?	___	___	___
	Are containers holding hazardous waste closed during storage, except when waste is added or removed? (If no, attach narrative.)	___	___	___
Subpart I - Containers §§264/265.173(a),(b) Management of Containers	Check here if containers are present at the facility. If no, go to Subpart J.	___	___	___
	Does the facility have procedures to ensure that containers holding hazardous waste are not opened, handled, or stored in a manner that may rupture the container or cause it to leak?	___	___	___
§§264/265.177 Special Requirements for Incompatible Wastes	Does the facility have procedures to ensure that incompatible wastes are not placed in the same containers or in unwashed containers that previously held incompatible waste?	___	___	___

Regulatory Citation/Description	RCRA Checklist Item	Yes	No	Potential Violation
	Are storage containers holding a hazardous waste that is incompatible with waste or other materials stored in nearby containers, piles, open tanks, or surface impoundments, separated from the other materials or protected from them by means of a dike, berm, wall, or other device?	___	___	_____
Subpart J - Tanks §§264/265.198(a)(1 and 2) Special Requirements for Ignitable or Reactive Wastes	Are ignitable or reactive wastes treated, rendered, or mixed before or immediately after placement in the tank so that:			
	• The resulting mixture no longer meets the definition of an ignitable or reactive waste?	___	___	_____
	• Section 264.17(b) is complied with?	___	___	_____
	Are wastes stored or treated in such a way that they are protected from any material or conditions that may cause the wastes to react or ignite?	___	___	_____
	Note: Facilities do not need to comply with the above if the tank system is used for emergency purpose only.	___	___	_____
§§264/265.199(b) Incompatible Wastes	Before a hazardous waste is stored in a tank that previously held an incompatible waste or material, is it thoroughly decontaminated?	___	___	_____
§265.200(a)(b) Waste Analysis (Specific requirement for interim status facilities)	Is a waste analysis or trial treatment conducted whenever a tank system is used to store a hazardous waste substantially different from waste previously treated or stored; or used to treat chemically a hazardous waste with a substantially different process than any previously used in that system?	___	___	_____

Regulatory Citation/Description	RCRA Checklist Item	Yes	No	Potential Violation
	If no to §265.200(a):			
	Has written, documented information on similar waste under similar operating conditions been obtained to show that the proposed treatment or storage will meet the requirements of §265.194(a)?	___	___	___
Subpart O - Incinerators §§264/265.347(b) Monitoring and Inspections	Is the complete incinerator and associated equipment (pumps, valves, etc.) inspected daily for leaks, spills, and fugitive emissions?	___	___	___
§264.347(c) (Not applicable to interim status facilities)	Are emergency waste cut-off systems and associated alarms tested weekly?	___	___	___
§264.345(d) (Not applicable to interim status facilities)	Is the incinerator combustion zone sealed?	___	___	___
	If this is a rotary kiln incinerator, is there black smoke or evidence of emissions?	___	___	___
	Is the combustion zone pressure lower than atmospheric pressure? If no, what is the reading? (Explain below.)	___	___	___

If the pressure is not measured in the combustion zone, what alternative methods are used equivalent to maintenance of combustion zone pressure? (Explain below.)

Regulatory Citation/Description	RCRA Checklist Item	Yes	No	Potential Violation
§264.345(e) Operating Requirements (Not applicable to interim status facilities)	Determine whether there is a functioning system to automatically cut off waste feed to the incinerator when operating conditions deviate from the permitted levels? (Optional: Facilities can simulate operating conditions to trigger the shut-off ...inspector should observe actual shut-off)	___	___	_____
Part 264/265 Unplanned Incinerator Stack Emissions	How many times did the emergency bypass stack open during the past 6 months of operation?	___ Times		
	How long did it last each time in average?	___ Minutes		
	How many times was the automatic waste feed cut-off system activated during the past 30 days of operation?	___ Times		
	Due to CO excursion?	___ Times		
	Due to Temperature excursions?	___ Times		
	Due to Waste feed excursions?	___ Times		
	Other causes?	___ Times		

8. Land Disposal Restrictions Checklist

I. GENERAL INFORMATION

Facility: _____

U.S. EPA ID No.: _____

Street: _____

City: _____ State: _____ Zip: _____

Telephone: _____

Inspection Date: _____/____/____ Time: _____(am/pm)

Weather Conditions: _____

	Name	Agency/Title	Telephone
Inspectors:	_____	_____	_____
	_____	_____	_____
Facility Representatives:	_____	_____	_____
	_____	_____	_____

See Appendix B to determine which of the following LDR waste codes exist in the following categories at the facility:

	Generate	Transport	Treat	Store	Dispose

F001-F005 Solvents

_____

_____

_____

F020-F023
and F026-F028

_____

_____

_____

California List

_____

_____

_____

First Third
(40 CFR 268.101)

_____

_____

_____

Second Third
(40 CFR 268.11)

_____

_____

_____

Third Third
(40 CFR 268.12)

_____

_____

_____

INSPECTION SUMMARY

Processes That Generate LDR Wastes:

LDR Waste Management:

Summary:

Signature:

II.   WASTE IDENTIFICATION

A.   List waste codes which the facility handles in each of the following LDR categories*:

1.   F001 through F005 spent solvents (§268.30):

   _____

2.   F020-F023 and F026-F028 dioxin-containing wastes (§268.31):

   _____

3.   California List Wastes (§268.32) (See Appendix A):

   _____

4.   First Third Wastes (§268.33):

   _____

5.   Second Third Wastes (§268.34):

   _____

6.   Third Third Wastes (§268.35)**:

   _____

*See Appendix B.
** Note: Effective 09/25/90, large quantity generators and TSDs were required to use the toxicity characteristic leaching procedure (TCLP) instead of the extraction procedure (EP) for determining the toxicity characteristic (TC). Small quantity generators must comply with this new requirement by 03/29/91. The LDR applies to EP toxic wastes and not to "newly identified" wastes which exhibit TC. TC wastes will be regulated under 40 CFR Part 268 only after they have been designated with LDR treatment codes or standards. Even if the former EP waste is also TC characteristic, it is only subject to LDR if it continues to fail the EP. If it fails the TC, it will not be subject to LDR until EPA has established a new treatment standard.

B.   Waste Code Determination

1.   Have all waste codes been correctly identified for purposes of compliance with §§268.9 and 262.11.*

      Yes____   No____

   If no, list below:

   <u>Assigned Classification</u>   <u>Correct Classification</u>

* Areas of concern (See Section 2.0) include: California List/waste categories with more stringent treatment standards; listed/characteristic; multi-source/single-source leachate; P and U waste codes/F and K wastes; and waste code carry through principle.

Comments:_____

2.   Have both the listed and characteristic waste code been assigned, where a listed waste exhibits a characteristic? (§268.9(a))

Yes ___   No ___   NA ___

Comments:_____

3.   Is the waste classified as D001, non-High TOC Ignitable Liquids Subcategory? If so, has the waste been treated by INCIN, FSUBS, or RORGS? Has the generator determined the underlying hazardous constituents (as defined in §268.2) that are reasonably expected to be present in the waste? (§268.9(a))

Yes ___   No ___   NA ___

Comments:_____

4.   Is the waste classified as D002, and prohibited under §268.37? Has the generator determined the underlying hazardous constituents (as defined in §268.2) that are reasonably expected to be present in the waste?

Yes ___   No ___   NA ___

Comments:_____

5.   Has multi-source leachate been assigned the F039 waste code?* (§261.31)

Yes ___   No ___   NA ___

* Leachate derived exclusively from F020-F023 and/or F026-F028 dioxin wastes retains the individual waste codes.

If yes, was single-source leachate combined to form multi-source leachate? (55 FR 22623)

Yes ___   No ___

Comments:_____

C.   **Does the facility handle the following wastes (national capacity variances)?**

1.   Debris contaminated with wastes that had treatment standards set in the Third Third rule based on incineration, mercury retorting, or vitrification. See Appendix A (expires - 05/08/94). (§268.5)

     Yes ___   No ___   List_____

2.   Inorganic solid debris as defined in §268.2(g)*; includes chromium refactory bricks carrying EPA Hazardous Waste Nos. K048-K052 (expires - 05/08/94). (§268.35(c))

     Yes ___   No ___   List_____

     *Note: Incorrect reference (§268.2(a)(7)) in Third Third rule.

3.   Debris contaminated with wastes listed in §268.12, and/or debris contaminated with any characteristic wastes for which treatment standards are established in Subpart D of Part 268 (expires 05/08/94). (§268.35(e)(1))*

     Yes ___   No ___   List_____

     *Note: Generator must demonstrate a good faith effort to locate treatment capacity suitable for its waste, utilize such capacity if found available, or file a report as required by §268.5(g) by August 12, 1993 or within 90 days after the hazardous waste is generated (whichever is later) describing the generator's efforts to locate treatment capacity.

4.   Mixed radioactive hazardous debris contaminated with wastes listed in §268.12 and mixed with radioactive hazardous debris contaminated with any characteristic waste for which treatment standards are established in 40 CFR Part 268, Subpart D (expires 05/08/94). (§268.35(e)(2))*

     Yes ___   No ___   List_____

     *Note: Generator must demonstrate a good faith effort to locate treatment capacity suitable for its waste, utilize such capacity if found available, or file a report as required by §268.5(g) by August 12, 1993 or within 90 days after the hazardous waste is generated (whichever is later) describing the generator's efforts to locate treatment capacity.

III. GENERATOR REQUIREMENTS

A. **Wastewater/Non Wastewater Category and Treatability Group/Treatment Standard**

Identification*

> ***Note:** This information is generally available on LDR notifications. If not, waste
> profile data and other documentation should be checked.

1. F001-F005 Spent Solvent Wastes: Does the generator correctly determine the appropriate Wastewater/Non Wastewater Category and treatment standard for each F-solvent?

   Yes ___   No ___   NA ___

   If available, list each waste code and check the correct treatability group.

Waste Code	Wastewater*	Non Wastewater
_____	_____	_____
_____	_____	_____
_____	_____	_____

   Comments _____

   * Less than 1 percent by weight total organic carbon (TOC), or less than 1 percent by weight total F001- F005 solvent constituents listed in §268.41, Table CCWE. (§268.2(f)(1))

2. F020-F023 and F026-F028 Dioxin Wastes: Does the generator correctly determine the appropriate treatability group/treatment standard for each dioxin waste?

   Yes ___   No ___   NA ___

   If yes, list each waste code and check the correct treatability group.

Waste Code	Wastewater*	Non Wastewater
_____	_____	_____
_____	_____	_____
_____	_____	_____

   Comments _____

   * Less than 1 percent TOC by weight and less than 1 percent total suspended solids (TSS) by weight. (§268.2(f))

3. First, Second, and Third Third Wastes:

   a. Does the generator correctly determine the appropriate treatability group/treatment standard for each waste?

      Yes ___   No ___   NA ___

If available, list each waste code and check the correct treatability group:

Waste Code	Subcategory	Wastewater*	Non Wastewater
_____	_____	_____	_____
_____	_____	_____	_____
_____	_____	_____	_____

* Less than 1 percent TOC by weight <u>and</u> less than 1 percent total suspended solids (TSS) with the following exceptions: K011, K013, and K014 wastewaters (as generated) -- less than 5 percent by weight TOC and less than 1 percent by weight TSS; K103 and K104 wastewaters (as generated) - less than 4 percent by weight TOC and less than 1 percent by weight TSS. (§268.2(f)(2) and (3))

Comments _____

b.   Do the assigned treatment standards for listed wastes cover constituents that may cause the waste to exhibit any characteristics? (§268.9 (b))

     Yes ___   No ___   NA ___

c.   Does the generator specify alternative treatment standards for lab packs?* (268.42(c))

     Yes ___   No ___   NA ___

*Use of the alternative treatment standards is not required. (55 FR 22629)

If yes, do lab packs only contain the following wastes?* (§268.42(c)(2))

___ Organometallics: Part 268, Appendix IV constituents
___ Organics: Part 268, Appendix V constituents

* Unregulated wastes and hazardous wastes which meet treatment standards may be commingled in the appropriate Appendix IV and V lab pack. (55 FR 22629)

d.   Does the generator specify the treatment standards for the relevant F039 multi-source leachate constituents?*

     Yes ___   No ___   NA ___

*Use of the alternative treatment standards is required. (55 FR 22619)

4.   California List Wastes: Has the generator correctly identified the wastewater/non wastewater category and treatment standard/prohibition level for the following wastes? (55 FR 22675)

     a.   Liquid hazardous wastes containing PCBs ≥50 ppm (268.32(a)(2))

          Yes ___   No ___   NA ___

If yes, check the appropriate category (treatability group) (see §268.42(a)(1)):

    ___ 50 to 500 ppm PCBs
    ___ ≥500 ppm PCBs

b.   Listed or characteristic wastes containing ≥1,000 mg/l (liquids) or ≥1,000 mg/kg HOCs, (non-liquids), which are not declared hazardous by the HOC content (55 FR 22675)

        Yes ___   No ___   NA ___

If yes, check the appropriate category (see §268.42(a)(2)):

___   Dilute HOC wastewater with 1,000 mg/l to 10,000 mg/l HOCs) (268.32(a)(3))
___   All other HOCs greater than or equal to the prohibition level of 1,000 mg/l (liquids) or mg/kg (non-liquids) (268.32(e)(1) and (2))

c.   Liquid hazardous wastes that exhibit a characteristic and also contain ≥134 mg/l nickel and/or ≥130 mg/l thallium (55 FR 22675)

        Yes ___   No ___   NA ___

5.   National Capacity Variance Wastes: Have all applicable California List prohibitions been identified for wastes covered under national capacity variances? (See Appendix A.)

    Yes ___   No ___   ·NA ___

If a waste stream contains a mixture of wastes, and a variance only applies to some of the waste codes, has the generator identified all applicable treatment standards and California List prohibitions? (See Appendix A.)

    Yes ___   No ___   NA ·___

If California List prohibitions apply to wastestreams managed by the generator, complete the following table for each waste code, noting the date on which relevant national capacity variances expire.

Waste Code	Cal. List Applicability	Expiration Date
_____	_____	___/___/___
_____	_____	___/___/___
_____	_____	___/___/___

Comments _____

6. Treatment standards expressed as required technologies: Has the generator specified an alternative method to that required in §268.42?

   Yes ___   No ___   NA ___

   If yes, list the waste code, the technology specified in §268.42, the alternative method, and documentation of approval. (§268.42(b))

Waste Code	Required Technology	Alternative Method	Approval
_____	_____	_____	_____
_____	_____	_____	_____
_____	_____	_____	_____

   Comments _____

7. Does the generator mix multiple restricted wastes all containing a common constituent of concern, but which have different treatment standards?

   Yes ___   No ___

   If yes, did the generator select the most stringent treatment standards? (§§268.41(b) and 268.43(b))

   Yes ___   No ___

   Comments _____

B. **Waste Analysis**

   1. Does the generator determine whether restricted wastes exceed treatment standards/prohibition levels at the point of generation?* (§268.7(a) (53 FR 31208))

      Yes ___   No ___

      * Note: This determination may be made at the point of disposal if the waste only has a prohibition level in effect (52 FR 25765).

      If no, does the generator ship all restricted wastes as not meeting treatment standards?

      Yes ___   No ___

      Comments _____

   2. Which of the following analytical methods does the generator employ?*

      *Note: A "No" answer to applicable questions b through d does not necessarily constitute a violation. However, knowledge of waste is rarely adequate if a generator certifies that treatment standard criteria have been met.

a. Knowledge of waste:

   Yes ___   No ___

If yes, list the wastes for which applied knowledge was used and describe the basis
of determination. Attach documentation. (§268.7(a)(5))

_____

_____

b. TCLP*: Are wastes with treatment standards specified in §268.41 analyzed
    using TCLP?** (BDAT*** = stabilization/immobilization technology)

   Yes ___   No ___   · NA ___

*TCLP =   Toxicity Characteristic Leaching Procedure (Part 268, Appendix I,
          EPA Test Method 1311).

**See Section 268.40(a) for options for using TCLP or EP test methods.
***BDAT = best demonstrated available technology. See Appendix A.

If yes, list the wastes for which TCLP was used and provide the date of last
test, identify the frequency of testing, and note any problems. Attach test results.
(§268.7(a)(5))

_____

_____

c. Total constituent analysis: Are wastes with treatment standards specified
    in §268.43 analyzed using total constituent analysis?* (BDAT =
    destruction/removal technology)

   Yes ___   No ___   NA ___

*See Appendix C for exceptions.
If yes, list the wastes for which total constituent analysis was used and provide the
date of last test, identify the frequency of testing, and note any problems. Attach test
results. (§268.7(a)(5))

_____

_____

d.   Is the paint filter liquids test (PFELT) used to determine if California List wastes are *liquid* hazardous wastes?

Yes ___   No ___   NA ___

*PFLT = Paint Filter Liquids Test (Test Method 9095, EPA Publication No. SW-846)

If yes, list the wastes for which PFLT was used and provide the date of last test, identify the frequency of testing, and note any problems. Attach test results. (§268.7 (a)(5))

_____

_____

3.   Does the generator treat restricted wastes in 90-day tanks or containers regulated under §262.34 (permissible in some states)?

Yes ___   No ___   (If No, go to 4.)

Does the generator treat the wastes to meet appropriate treatment standards/prohibition levels?

Yes ___   No ___

If yes, has the generator prepared a waste analysis plan detailing the frequency of testing to be conducted? (§268.7(a)(4))

Yes ___   No ___   (If No, go to 4.)

Does the plan fulfill the following? (§268.7(a)(4)(i))

___   Based on a detailed chemical and physical analysis of a representative sample
___   Contains information necessary to treat the wastes in accordance with Part 268 requirements

Has the plan been filed with the Regional Administrator (return receipt, Federal Express slip, etc. required for verification)? (§268.7(a)(4)(ii))

Yes ___   No ___

Comments _____

_____

4.   Dilution Prohibition (§268.3):

a.   Does the generator mix prohibited* wastes with different treatment standards?

*See Appendix C for distinction between restricted and prohibited wastes.

Yes ___   No ___   (If No, go to b.)

List the wastes_____

Are the wastes amenable to the same type of treatment? (55 FR 22666)

Yes ___   No ___

Comments_____

b.   Does the generator dilute prohibited wastes to meet treatment standard criteria, or render them non-hazardous? (55 FR 22665-22666)

Yes ___   No _⌐_   (If No, go to c.)

Check appropriate category:

___ Dilutes to meet treatment standards
___ Dilutes to render waste non-hazardous

Do the wastes fall into the following categories? (Check if appropriate.)

___   Characteristic wastes managed in treatment systems regulated under the Clean Water Act (§268.3(b)), (55 FR 22665)
___   Treatment standard specified in §§268.41 or 268.43

If the wastes do not fall into the above categories, briefly describe the conditions under which they were diluted.

_____

c.   Based on an assessment of points a and b, and any other relevant circumstances, does the generator dilute prohibited wastes as a substitute for adequate treatment? (§268.3(a))

Yes ___   No ___

Comments _____   .

5. F039 Multi-source leachate: Has the generator run an initial analysis for all constituents of concern in §§268.41 and 268.43? (55 FR 22620)

Yes ___   No ___   NA ___

C. Management

1. On-Site Management

a. Are restricted wastes treated or (other than in a RCRA exempt unit) stored for greater than 90 or 180 days, or disposed on site?

Yes ___   No ___

(If yes, the TSD Checklist must also be completed.)

Comments_____

b. If the generator treats characteristic wastes in systems regulated under the Clean Water Act, have the following been documented: the determination of restriction, how restricted wastes are managed, and why wastes discharged pursuant to an NPDES permit are not prohibited (if applicable)? (§268.7(a)(6)) (55 FR 22662)

Yes ___   No ___   NA ___

c. If the generator treats characteristic wastes in RCRA exempt units to render them non-hazardous, are the wastes managed as restricted prior to entering the exempt unit (§268.7(a)(6)) until the applicable treatment standards are met?* (§268.9(d))

Yes ___   No ___   NA ___

* This applies to both concentration based treatment standards specified in §§268.41 and 268.43, and to some §268.42 required methods which result in treatment below the characteristic level. See Appendix D.

d. If a waste is excluded from regulation or from the definition of solid or hazardous waste subsequent to the point of generation, does the generator comply with the requirements of §268.7(a)(6) (56 FR 3866-3867)? If the generator determines that he is managing a restricted waste that is excluded from the definition of hazardous or solid waste or exempt from Subtitle C regulation, under §§261.2-261.6 subsequent to the point of generation, is there a one-time notice in the facility's file stating such generation, subsequent exclusion from the definition of hazardous or solid waste or exemption from Subtitle C regulation, and the disposition of the waste?

2. Off-Site Management: Waste Exceeds Treatment Standards

    a. Does the generator ship any waste that exceeds treatment standards/ prohibition levels (not subject to a national capacity variance) to an off-site treatment or storage facility?

        Yes ___   No ___   (If No, go to 3.)

        Identify waste code(s) and off-site treatment or storage facilities to which wastes are shipped.

Waste Code	Receiving Facility

        Does the generator provide a notification to the treatment or storage facility? (§268.7(a)(1))

        Yes ___   No ___   (If No, go to 3.)

        If the generator specifies alternative treatment standards for lab packs, is the certification required in §268.7(a)(8) or (9) included with the notification?

        Yes ___   No ___   NA ___

    b. Is a notification sent with each waste shipment?

        Yes ___   No ___

        If no, is the waste subject to a tolling agreement pursuant to §62.20(e) (small quantity generator only)?

        Yes ___   No ___   (If No, go to 3.)

        List waste codes and subsequent handler with whom a contractual tolling agreement is held.

Waste Code	Subsequent Handler

        Did the small quantity generator provide a notification to the receiving facility with the first waste shipment subject to the tolling agreement? (§268.7(a)(10))

        Yes ___   No ___

3.   Off-Site Management: Waste Meets Treatment Standards

  a.   Does the generator ship waste that meets treatment standards/prohibition
        levels to an off-site disposal facility?

        Yes ___   No ___   (If No, go to 4.)

        Identify waste code(s) and off-site disposal facilities:

          Waste Code   Receiving Facility
          _____   _____
          _____   _____
          _____   _____

        Does the generator provide a notification and a certification to the disposal
        facility? (§268.7(a)(2)(i) and (ii))?

        Yes ___   No ___: (If No, go to d.)

  b.   Are a notification and a certification sent with each waste shipment?

        Yes ___   No ___

        If no, is the waste subject to a tolling agreement pursuant to §262.20(e)
        (small quantity generator only)?

        Yes ___   No ___   (If No, go to c.)

        List waste codes and subsequent handler with whom a contractual
        tolling agreement is held.

          Waste Code   Subsequent Handler
          _____   _____
          _____   _____
          _____   _____

        Did the small quantity generator provide a notification and a certification
        to the receiving facility with the first waste shipment subject to the tolling
        agreement? (§268.7(a)(10))

        Yes ___   No ___

  c.   Are characteristic wastes which have been rendered non-hazardous shipped
        to a Subtitle D facility?

        Yes ___   No ___   NA ___   (If No or NA, go to 4.)

Complete the following table:

Waste Code	Receiving Facility
_____	_____
_____	_____
_____	_____

Are a notification and a certification for each shipment sent to the Regional Administrator or authorized State? (§§268.9(d)(1) and 268.7(b)(5))?

Yes ___   No ___

4.   Off-Site Management: Wastes Subject to Variances, Extensions, or Petitions

a.   Does the generator ship wastes to a treatment, storage, or disposal facility which are subject to a national capacity variance (Part 268, Subpart C), or case-by-case extension (§268.5)?

Yes ___   No ___   (If No, go to 5.)

Complete the following table:

Waste Code	Receiving Facility
_____	_____
_____	_____
_____	_____

Does the generator provide notification to the off-site receiving facility that the waste is not prohibited from land disposal? (§268.7(a)(3))

Yes ___   No ___

b.   Is a notification sent with each waste shipment?

Yes ___   No ___

If no, is the waste subject to a tolling agreement pursuant to §262.20(e) (small quantity generator only)?

Yes ___   No ___   (If No, go to 5.)

List waste codes and subsequent handler with whom a contractual tolling agreement is held.

Waste Code	Subsequent Handler
_____	_____
_____	_____
_____	_____

Did the small quantity generator provide a notification to the receiving facility with the first waste shipment subject to the tolling agreement? (§268.7(a)(10))

Yes ___   No ___

5.   Records Retention

Does the generator retain on site copies of all notifications, certifications, and other relevant documents for a period of 5 years? (§268.7(a)(7))

Yes ___   No ___

Are copies of relevant tolling agreements, along with the LDR notification and/or certification, kept on site for at least 3 years after expiration or termination of the agreement? (§268.9)

Yes ___   No ___   NA ___

Do LDR documents reflect proper management of wastes previously covered under expired national capacity variances, case by case extensions and the soft hammer provision*?

Yes ___   No ___   NA ___

* See Appendix B. Note that the soft hammer provision expired as of 05/08/90. Soft hammer wastes which had treatment standards established in the Third Third rule were granted a minimum 90-day national capacity variance to 08/08/90.

Comments_____

D.   Treatment Using RCRA 40 CFR Parts 264 and 265 Exempt Units or Processes

1.   Are restricted wastes-treated in RCRA exempt units (e.g., distillation units, wastewater treatment tanks, elementary neutralization, etc.)?

Yes ___   No ___   (If No, do not complete this section.)

List types of waste treatment units and processes:

Waste Code	Type of Treatment	Treatment Units and Processes

2.   Are treatment residuals generated from these units?

Yes ___   No ___

Comments_____

3. Are residuals further treated, stored for greater than 90/180 days, or disposed on site?

    Yes ___   No ___   NA ___

    (If yes, the TSD checklist must also be completed.)

E. Additional Comments, Concerns, or Issues Not Addressed in the Checklist:

_____
_____
_____
_____
_____
_____
_____
_____
_____
_____
_____
_____
_____
_____
_____
_____
_____
_____
_____
_____
_____
_____
_____
_____

APPENDIXES

Appendix A

U.S. EPA Regional Offices

EPA Region I

Connecticut, Massachusetts, Maine, New Hampshire, Rhode Island, Vermont

JFK Federal Building
Boston, MA 02203
(617) 565-3420 (General Information)
(617) 573-5758 (Hazardous Waste Ombudsman)

EPA Region II

New Jersey, New York, Puerto Rico, Virgin Islands

26 Federal Plaza
New York, NY 10278
(212) 264-2657 (General Information)
(212) 264-2980 (Hazardous Waste Ombudsman)

EPA Region III

Delaware, Maryland, Pennsylvania, Virginia, West Virginia, District of Columbia

841 Chestnut Street
Philadelphia, PA 19107
(215) 597-9800 (General Information)
(215) 597-2842 (Hazardous Waste Ombudsman)

EPA Region IV

Alabama, Florida, Georgia, Kentucky, Mississippi, North Carolina, South Carolina, Tennessee

345 Courtland Street, N.E.
Atlanta, GA 30365
(404) 347-4727 (General Information)
(404) 347-3004 (Hazardous Waste Ombudsman)

EPA Region V

Illinois, Indiana, Michigan, Minnesota,
Ohio, Wisconsin

77 West Jackson Boulevard
Chicago, IL 60604
(312) 353-2000 (General Information)
(312) 886-0981 (Hazardous Waste Ombudsman)

EPA Region VI

Arkansas, Louisiana, New Mexico, Oklahoma, Texas

1445 Ross Avenue
Dallas, TX 75270
(214) 655-6444 (General Information)
(214) 655-8527 (Hazardous Waste Ombudsman)

EPA Region VII

Iowa, Kansas, Missouri, Nebraska

726 Minnesota Avenue
Kansas City, KS 66101
(913) 551-7000 (General Information)
(913) 551-7050 (Hazardous Waste Ombudsman)

EPA Region VIII

Colorado, Montana, North Dakota, South Dakota,
Utah, Wyoming

One Denver Place
999 18th Street
Denver, CO 80202
(303) 293-1603 (General Information)
(303) 294-1111 (Hazardous Waste Ombudsman)

EPA Region IX

Arizona, California, Hawaii, Nevada, American Samoa,
Guam, Trust Territories of the Pacific

75 Hawthorne Street
San Francisco, CA 94105
(415) 744-1305 (General Information)
(415) 744-2124 (Hazardous Waste Ombudsman)

EPA Region X

Alaska, Idaho, Oregon, Washington

1200 Sixth Avenue
Seattle, WA 98101
(206) 553-4973 (General Information)
(206) 553-6901 (Hazardous Waste Ombudsman)

Appendix B

State Environmental Agency Offices

Alabama

Alabama Department of Environmental Management
1751 Federal Drive
P.O. Box 301463
Montgomery, AL 36130-1463

Alaska

Alaska Department of Environmental Conservation
Division of Environmental Quality
P.O. Box #O
Juneau, AK 99801

American Samoa

Environmental Quality Commission
Government of American Samoa
Pago Pago, American Samoa 96799

Arizona

Arizona Department of Environmental Quality
2005 N. Central Avenue
Phoenix, AZ 85004

Arkansas

Arkansas Department of Pollution Control and Ecology
P.O. Box 9583
Little Rock, AR 72219

California

California Environmental Protection Agency
1020 9th Street
Sacramento, CA 95814

Colorado

Colorado Department of Health
4300 Cherry Creek Drive South
Denver, CO   80222

Connecticut

Department of Environmental Protection
State Office Building
165 Capitol Avenue
Hartford, CT   06106

Delaware

Department of Natural Resources and Environmental Control
P.O. Box 1401
89 Kings Highway
Dover, DE   19903

District of Columbia

D.C. Department of Consumer and Regulatory Affairs
Environmental Control Division
2100 Martin Luther King, Jr. Avenue, S.E., Suite 203
Washington, DC   20020

Florida

Department of Environmental Regulation
Twin Towers Office Building
2600 Blair Stone Road
Tallahassee, FL   32301

Georgia

Georgia Department of Natural Resources
Floyd Towers East
205 Butler Street, S.E.
Atlanta, GA   30334

Guam

Guam Environmental Protection Agency
P.O. Box 2999
Agana, Guam   96910

Hawaii

Hawaii Department of Health
P.O. Box 3378
Honolulu, HI 96801

Idaho

Idaho Department of Health and Welfare
Idaho State House
450 W. State Street
Boise, ID 83720

Illinois

Illinois Environmental Protection Agency
2200 Churchill Road
Springfield, IL 62706

Indiana

Indiana Department of Environmental Management
105 S. Meridian Street
P.O. Box 6015
Indianapolis, IN 46225

Iowa

Iowa Department of Natural Resources
900 East Grand Avenue
Henry A. Wallace Building
Des Moines, IA 50319-0034

Kansas

Kansas Department of Health and Environment
Forbes Field, Building 321
Topeka, KS 66620

Kentucky

Department of Environmental Protection
Fort Boone Plaza
18 Riley Road
Frankfort, KY 40601

Louisiana

Louisiana Department of Environmental Quality
P.O. Box 44307
625 N. 4th Street
Baton Rouge, LA   70804

Maine

Department of Environmental Protection
State House Station #17
Augusta, ME   04333

Maryland

Maryland Department of the Environment
201 W. Preston Street
Baltimore, MD   21201

Massachusetts

Massachusetts Department of Environmental Protection
One Winter Street, 5th FL
Boston, MA   02108

Michigan

Environmental Protection Bureau
Department of Natural Resources
Box 30038
Lansing, MI   48909

Minnesota

Minnesota Pollution Control Agency
520 Lafayette Road, North
St. Paul, MN   55155

Mississippi

Department of Natural Resources
P.O. Box 10385
Jackson, MS   39209

Missouri

Department of Natural Resources
Jefferson Building - 205 Jefferson St.
P.O. Box 176
Jefferson City, MO   65102

Montana

Department of Health and Environmental Sciences
Cogswell Building
Helena, MT 59620

Nebraska

Department of Environmental Control
State House Station
P.O. Box 94877
Lincoln, NE 68509

Nevada

Department of Conservation and Natural Resources
Capitol Complex
201 South Fall Street
Carson City, NV 89710

New Hampshire

Department of Health and Welfare
Health and Welfare Building
6 Hazen Drive
Concord, NH 03301

New Jersey

Department of Environmental Protection
401 East State Street
CN 028
Trenton, NJ 08625

New Mexico

New Mexico Health and Environment Department
P.O. Box 968
Santa Fe, NM 87504-0968

New York

Department of Environmental Conservation
50 Wolfe Road
Albany, NY 12233

North Carolina

Department of Environment, Health and Natural Resources
Division of Environmental Management
P.O. Box 2091
Raleigh, NC 27602

North Dakota

North Dakota State Department of Health
1200 Missouri Avenue
Box 5520
Bismarck, ND 58502-5520

Northern Mariana Islands

Department of Public Health and Environmental Services
Commonwealth of the Northern Mariana Islands
Office of the Governor
Saipan, Mariana Islands 96950

Ohio

Ohio Environmental Protection Agency
1800 Watermark Drive
P.O. Box 1049
Columbus, OH 43266-0149

Oklahoma

Oklahoma Department of Environmental Quality
P.O. Box 53551
1000 Northeast 10th Street
Oklahoma City, OK 73152

Oregon

Department of Environmental Quality
811 Southeast 6th Avenue
Portland, OR 97204

Pennsylvania

Pennsylvania Department of Environmental Resources
P.O. Box 2063
Fulton Building
Harrisburg, PA 17120

Puerto Rico

Environmental Quality Board
Santurce, PR 00910-1488

Rhode Island

Department of Environmental Management
204 Cannon Building
75 Davis Street
Providence, RI 02908

South Carolina

Department of Health and Environmental Control
2600 Bull Street
Columbia, Sc 29201

South Dakota

Department of Water and Natural Resources
523 East Capitol
Foss Building
Pierre, SD 57501

Tennessee

Tennessee Department of Environment and Conservation
L & C Tower
401 Church Street
Nashville, TN 37243

Texas

Texas Water Commission
P.O. Box 13087, Capitol Station
Austin, TX 78711-3087

Utah

Utah Department of Environmental Quality
P.O. Box 16700
288 North, 1460 West Street
Salt Lake City, UT 84116-0700

Vermont

Vermont Department of Environmental Conservation
103 South Main Street
Waterbury, VT 05676

Virgin Islands

Department of Conservation and Cultural Affairs
P.O. Box 4399, Charlotte
St. Thomas, VI   00801

Virginia

Department of Waste Management
Monroe Building, 11th FL
101 North 14th Street
Richmond, VA   23219

Washington

Washington Department of Ecology
Mail Stop PV-11
Olympia, WA   98504

West Virginia

West Virgina Department of Natural Resources
1260 Greenbriar Street
Charleston, WV   25311

Wisconsin

Department of Natural Resources
P.O. Box 7921
Madison, WI   53707

Wyoming

Department of Environmental Quality
122 West 25th Street
Herschler Building
Cheyenne, WY   82002

Index

About the Author

Mark Dennison is an attorney and author or co-author of numerous books and articles dealing with environmental law and regulatory compliance issues. His books include *OSHA and EPA Process Safety Management Requirements: A Practical Guide for Compliance* (VNR, 1994); *Hazardous Waste Regulation Handbook: A Practical Guide to RCRA and Superfund* (1994); *Understanding Solid and Hazardous Waste Identification and Classification* (1993); *Wetlands and Coastal Zone Regulation and Compliance*, with Steven Silverberg (1993); *Wetlands: Guide to Science, Law, and Technology*, with James Berry (1993). Mr. Dennison is also the editor-in-chief of the monthly newsletter, *Environmental Strategies for Real Estate*. He is in private practice in Ridgewood, New Jersey, where he specializes in environmental, land use, real estate, and zoning law. He is admitted to practice in New Jersey and New York. Mr. Dennison holds a B.A., *magna cum laude*, from the State University of New York (Oswego), an M.A. from Syracuse University, and a J.D. from New York Law School.

NOTES

NOTES

NOTES

NOTES